Robust Control of Robots

Adriano A. G. Siqueira · Marco H. Terra
Marcel Bergerman

Robust Control of Robots

Fault Tolerant Approaches

Springer

Dr. Adriano A. G. Siqueira
Engineering School of São Carlos
University of São Paulo
Av. Trabalhador Sãocarlense
400 São Carlos
SP 13566-590
Brazil
e-mail: siqueira@sc.usp.br

Dr. Marcel Bergerman
CMU Robotics Institute
5000 Forbes Avenue
Pittsburgh
PA 15213-3890
USA
e-mail: marcel@cmu.edu

Dr. Marco H. Terra
Engineering School of São Carlos
University of São Paulo
Av. Trabalhador Sãocarlense
400 São Carlos
SP 13566-590
Brazil
e-mail: terra@sc.usp.br

Additional material to this book can be downloaded from http://extra.springer.com

ISBN 978-1-4471-5792-2 ISBN 978-0-85729-898-0 (eBook)
DOI 10.1007/978-0-85729-898-0
Springer London Dordrecht Heidelberg New York

British Library Cataloguing in Publication Data
A catalogue record for this book is available from the British Library

Cover design: eStudio Calamar S.L.

Printed on acid-free paper

Springer is part of Springer Science+Business Media (www.springer.com)

To My Wife Flaviane and My Son João Vítor

A. A. G. S.

To My Son Diego

M. H. T.

To My Family

M. B.

Preface

The world is teeming with machines that grow crops, process our food, drive us to work, assemble products, clean our homes, and perform thousands of other daily tasks. They are complex systems built on a myriad of electronic, mechanic, and software components, each one prone to malfunction and even fail at any given time. Robust, fault tolerant control of such machines is key to guaranteeing their performance and avoiding accidents. Robotic manipulators, in particular, are especially important when it comes to robust, fault tolerant control. Our society relies on these machines for a large variety of industrial operations; any unscheduled downtime caused by a faulty component can have significant economic costs—not to mention the consequences of a potential injury.

Robust and fault tolerant systems have been studied extensively by academic and industrial researchers and many different design procedures have been developed in order to satisfy rigorous robustness criteria. An important class of robust control methods, introduced by G. Zames in 1980, is based on $\mathcal{H}_\infty$ theory. The main concept behind this approach is the robustness of the control system to internal uncertainties and exogenous disturbances. Hundreds of works were written extending the seminal results obtained by Prof. Zames. Some of them are sufficiently elegant and effective to be of value in industrial environments. Transforming theory into practice, however, is not a trivial task, as the mathematics involved in robust control can be daunting. This monograph proposes to bridge the gap between robust control theory and applications, with a special focus on robotic manipulators.

The book is organized in nine chapters. In Chap. 1 we present the experimental robot manipulator system used throughout the book to illustrate the various control methodologies discussed. We also present there the simulation and control environment we use to develop and test the methodologies. The environment, named CERob for Control Environment for Robots, is a freeware included with this book and available at http://extras.springer.com. The remaining eight chapters are divided in three parts. Part 1 (Chaps. 2–4) deals with robust control of regular, fully-actuated robotic manipulators. Part 2 (Chaps. 5–6) deals with robust fault tolerant control of robotic manipulators, especially the post-failure control problem. Finally, Part 3 (Chaps. 7–9) deals with robust control of cooperative robotic manipulators.

In Chaps. 2, 3, and 4 we present model-based linear, non-linear, adaptive, and neural network-based $\mathcal{H}_\infty$ controllers for robotic manipulators. Models based on the Lagrange–Euler formulation and neural networks are used to enable robust control of robots where performance, stability and convergence are guaranteed. One interesting scenario in robot modeling is when the neural network works as a complement of the Lagrange–Euler equations to decrease modeling errors. In these chapters we also explore the use of output feedback controllers, motivated by the fact that in some cases sensors are not available to measure the full array of variables needed for robot control.

In Chaps. 5 and 6 we present strategies to control the position of underactuated manipulators, or manipulators equipped with both regular (active) and failed (passive) joints based on linear parameter-varying models and linear matrix inequalities, and also on game theory. The objective in these chapters is to demonstrate that the system is able to reject disturbances while achieving good position tracking performance. For robotic systems subject to faults, we present a fault tolerant methodology based on linear systems subject to Markovian jumps. We describe in detail the design of $\mathcal{H}_2, \mathcal{H}_\infty$, and mixed $\mathcal{H}_2/\mathcal{H}_\infty$ trajectory-following controllers for manipulators subject to several consecutive faults.

In Chaps. 7, 8, and 9 we consider actuated and underactuated cooperative manipulators. One of the most important issues in the robust control of cooperative manipulators is the relationship between disturbance rejection and control of squeeze forces on the load, particularly when the manipulator is underactuated.

Throughout the book we illustrate the concepts presented with simulations and experiments conducted with two 3-link planar robotic manipulators especially designed to pose as fully-actuated or underactuated devices.

São Carlos, Brazil — Adriano A. G. Siqueira
São Carlos, Brazil — Marco H. Terra
Pittsburgh, USA — Marcel Bergerman

Acknowledgments

This book would not have been possible without the help and support of our colleagues and former students from the Mechatronics Laboratory and Intelligent Systems Laboratory at the University of São Paulo. In particular, we would like to express our gratitude to our research partner Prof. Renato Tinós for his contributions to the material presented in Chaps. 7 and 8. We also express our appreciation for the continued support of the Research Council of the State of São Paulo (FAPESP), under grants 98/0649-5, 99/10031-1, 01/12943-0, and the Brazilian National Research Council, under grant 481106/2004-9.

The material presented in this book was partially published previously in scientific journals. We would like to thank the publishers for permitting the reproduction of that content, especially: Cambridge University Press for the paper "Nonlinear $\mathcal{H}_\infty$ controllers for underactuated cooperative manipulators," by Adriano A. G. Siqueira and Marco H. Terra, published in Robotica, vol. 25, no. 4, in 2007; Elsevier for the papers "Nonlinear mixed control applied to manipulators via actuation redundancy," by Adriano A. G. Siqueira, Marco H. Terra, and Benedito C. O. Maciel, published in Control Engineering Practice, vol. 14, no. 4, in 2006; "A fault tolerance framework for cooperative robotic manipulators," by Renato Tinós, Marco H. Terra, and Marcel Bergerman, published in Control Engineering Practice, vol. 15, no. 5, in 2007; and "Neural network-based control for fully actuated and underactuated cooperative manipulators," by Adriano A. G. Siqueira and Marco H. Terra, published in Control Engineering Practice, vol. 17, no. 3, in 2009; IEEE for the papers "Nonlinear and Markovian controls of underactuated manipulators," by Adriano A. G. Siqueira and Marco H. Terra, published in IEEE Transactions on Control Systems Technology, vol. 12, no. 6, in 2004; "Motion and force control of cooperative robotic manipulators with passive joints," by Renato Tinós, Marco H. Terra, and João Y. Ishihara, published in IEEE Transactions on Control Systems Technology, vol. 14, no. 4, in 2006; and "A fault-tolerant manipulator robot based on $\mathcal{H}_2, \mathcal{H}_\infty$ and mixed $\mathcal{H}_2/\mathcal{H}_\infty$ Markovian controls," by Adriano A. G. Siqueira and Marco H. Terra, published in IEEE/ASME Transactions on Mechatronics, vol. 14, no. 2, in 2009.

Contents

Acronyms

AAA	Active-active-active
AAP	Active-active-passive
ADP	Adaptive controller
ANN	Artificial neural network
APA	Active-passive-active
APP	Active-passive-passive
CERob	Control environment for robots
CM	Center of mass
CMCE	Cooperative manipulator control environment
DC	Direct current
DLCC	Dynamic load carrying capacity
dll	Dynamically linked library
DOF	Degree of freedom
FDI	Fault detection and isolation
FSJF	Free-swinging joint fault
FTMCE	Fault tolerant manipulator control environment
GTH	Game theory controller
HBC	Hybrid controller
HIN	Linear $\mathcal{H}_\infty$ controller
JPF	Joint position fault
JVF	Joint velocity fault
LJF	Locked joint fault
LMI	Linear matrix inequality
LPV	Linear parameter-varying
MJLS	Markov jump linear systems
MLP	Multilayer perceptron
MTD	Mean time to detection
NET	Neural network controller
NLH	Nonlinear $\mathcal{H}_\infty$ controller
PAA	Passive-active-active
PAP	Passive-active-passive

PCI	Peripheral component interconnect
PD	Proportional-derivative
PPA	Passive-passive-active
RBFN	Radial basis function network
UARM	Underactuated robotic manipulator
UMCE	Underactuated manipulator control environment
VSC	Variable structure controller

Chapter 1
Experimental Set Up

1.1 Introduction

The field of Robotics is, by its very nature, an experimental one. No robot control methodology can be deemed to perform satisfactorily if it has not been validated on an actual physical system. In this book we illustrate all control methods presented by applying them to custom-designed robotic manipulators.

The manipulators are two 3-link open-chain, serial link arms built by Ben Brown, Jr. from Pittsburgh, PA, USA, which we name the UARMs, or UnderActuated Robotic Manipulators (Fig. 1.1). The two most salient features of these manipulators are that their joint motors possess very low friction and are equipped with on/off brakes, thus allowing us to simulate a variety of joint failure conditions. We created an open source MATLAB®-based UARM simulator that readers can utilize to validate the control methodologies presented throughout the chapters. The simulator includes also the Matlab source code for all methods described. This Control Environment for Robots (CERob) is in fact more than just a standard simulator; in our laboratory, control methodologies can be validated in the virtual manipulators and then transferred to the actual ones at the click of a button.

In the first part of this chapter we describe in detail the UARM hardware and its dynamic model. In the second part we describe the basic functionality of CERob. Specific details on CERob as it applies to particular controllers are presented in the pertinent chapters.

1.2 UARM Experimental Manipulator

1.2.1 Hardware

Each UARM is a 3-link planar, open-chain, serial-link manipulator. They are equipped with low-friction DC motors directly connected to the links, with no gearboxes. When the motors are powered, the joints behave as regular

A. A. G. Siqueira et al., *Robust Control of Robots*,
DOI: 10.1007/978-0-85729-898-0_1, © Springer-Verlag London Limited 2011

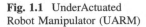

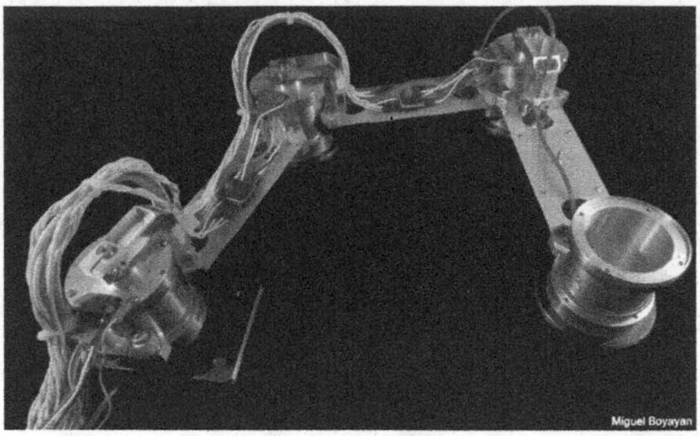

Fig. 1.1 UnderActuated Robot Manipulator (UARM)

fully-actuated ones; and when they are not powered, the joints can move essentially freely and dub as unactuated or passive joints. The manipulator's configuration (fully-actuated or underactuated) can then be defined at will in real time by simply powering or not each joint. Additionally, the joints are equipped with diaphragm-based on/off brakes, which can be used to simulate locking-type faults or to enable underactuated manipulator position control. Joints are numbered from 1 to 3, with joint 1 fixed to a smooth marble table. A dummy load can be attached to the end-effector's cup-shaped housing to provide for meaningful manipulation experiments. The entire system resides on a horizontal plane and runs on a thin air film that reduces to practically zero the friction with the table. Pressurized air pumps and computer-controlled solenoid valves complete the hardware of the system.

Incremental encoders with quadrature decoding, located at the top of each joint, are used to measure the relative angular joint positions. The angular velocities are computed via numerical differentiation and low-pass filtering. Such procedure is known to result in measurement noise and can therefore lead to poor position control performance. This is one of our motivations to use output feedback control laws, where only joint position measurements are used.

An equipment board provides a mounting surface for the motor amplifiers, valves and air pressure regulators for brake and flotation air, and a custom interface board. Two feeding voltages are supplied by a power unit: 48 V/20 A for the motors, and 24 V/1 A for the interface board. A kill switch ("emergency stop") mounted in a small yellow enclosure can be located remotely, and controls the power supply to the interface board and amplifiers. The equipment board also provides a location where the UARM can be secured for safe transport without disconnecting it from the rest of the system. Figure 1.2 shows the complete system, with the robot manipulator, the equipment board, the power supply unit, and the control computer.

The communication between the control computer and the hardware is performed by a PCI (Peripheral Component Interconnect) input–output board from Motenc. This device is able to control up to eight servo motors simultaneously

Fig. 1.2 Complete UARM hardware setup

(eight robot joints). The Motenc board connects to the custom interface board via two 50-conductor ribbon cables. The features available on this board include:

- 8 differential encoder inputs, 32-bit resolution;
- 8 analog outputs, ±10 V range, 13-bit resolution;
- 8 analog inputs, ±5 V range, 14-bit resolution;
- 100 digital I/O (68 inputs and 32 outputs) in four 50-pin headers, opto-22 compatible;
- +5 V available on headers, fused (resetable), max current 2 A;
- Programmable timer interrupts;
- Watchdog timer;
- Hardware board ID for multiple board applications;
- Filters at digital inputs to remove high frequency noise; and
- Hardware ESTOPs.

The designer can access the board I/O channels by using the *dll* library provided by the manufacturer. Table 1.1 shows the MATLAB®-based functions we developed to communicate with the UARM hardware. Most of the time only four commands are necessary to control the robot: those to set the desired voltages, read the current encoders' values, reset the encoders, and activate or release the brakes, at the same time activating or inhibiting the motors.

Table 1.1 MATLAB®-based functions for control of the UARM

Function	Description
set_dac_all_stg([v1 v2 v3])	Set the desired voltages to the DC motors
get_position	Read joint encoders
set_encoder_one_stg(enc, value)	Set encoder to the specified value
setbrakemotor(value)	Activate/release brakes and activate/inhibit motors

Electrical and mechanical details of the system are as follows:

1. Links
   - Link length: 20.3 cm
   - Arm length from center of joint 1 to center of tip: 60.96 cm
   - Joint size: 76 mm (diameter) × 86 mm (height)
   - Joint mass: 670 g
   - Tip mass: 220 g (default, customized by user)
   - Link mass: 30 g (excluding wires, air hoses, and connectors).
2. Joint motors
   - Model: Kollmorgen RBE-1213 brushless DC
   - Nominal voltage: 48 V CC
   - Winding resistance: 2.4 Ω
   - Torque constant: 0.14 Nm/A
   - Peak torque: 2.8 Nm
   - Back EMF constant: 15 V/kRPM
   - Continuous stall torque: 0.35 Nm
   - Motor mass: 344 g
   - Rotor inertia: 0.0000148 kg m^2.
3. Motor amplifiers
   - Model: Elmo SBA 10/100H-4
   - Peak current: 20 A
   - Continuous current: 10 A
   - Supply voltage: 20–90 V CC
   - Current-to-voltage constant (adjustable): 1.61 A/V.
4. Brakes
   - Type: air actuated diaphragm
   - Pressure: 100 psi (700 kPa) max.
   - Valves: Clippard model EVO-3M, 24 VDC, 0.67 W
   - Torque: 2.8 Nm max.
5. Optical encoders
   - Model: Hewlett Packard HEDS 9040-T00
   - Disk: HEDS 6140-T08
   - Lines: 2000/revolution
   - Counts: 8000/revolution after quadrature decoding.
6. Air bearings
   - Orifice diameter: 0.36 mm
   - Air gap: approx. 0.08 mm
   - Air supply pressure: 100 psi (700 kPa) max.

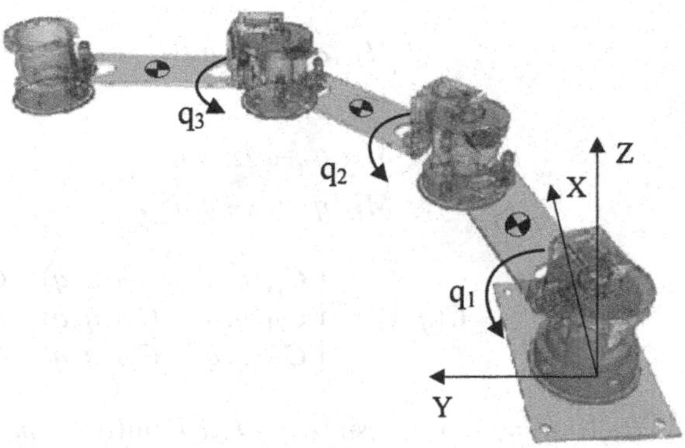

Fig. 1.3 Schematic representation of the UARM

1.2.2 Dynamic Model

Figure 1.3 shows a schematic representation of the UARM and the coordinate frames assigned to compute the direct and inverse kinematic models. The robot's dynamic equations, which all control laws considered in this book are based on, can be computed via Lagrange theory [1] and is given by:

$$\tau = M(q)\ddot{q} + C(q,\dot{q})\dot{q} + F(\dot{q}) + G(q), \tag{1.1}$$

where q is the n-dimensional joint position vector, $M(q)$ is the $n \times n$ symmetric positive definite inertia matrix, $C(q,\dot{q})$ is the $n \times n$ Coriolis and centripetal torque matrix, $F(\dot{q})$ is the n-dimensional velocity-dependent frictional torque vector, $G(q)$ is the n-dimensional gravitational torque vector and τ is the n-dimensional applied torque vector. Because the UARM moves in the horizontal plane, the gravitational term vanishes. The elements of the matrices in Eq. 1.1 are given by:

Dynamic Matrices of UARM:

$$M(q) = \begin{bmatrix} M_{11}(q) & M_{12}(q) & M_{13}(q) \\ M_{21}(q) & M_{22}(q) & M_{23}(q) \\ M_{31}(q) & M_{32}(q) & M_{33}(q) \end{bmatrix},$$

$$\begin{aligned}
M_{11}(q) =& I_1 + I_2 + I_3 + m_1 l_{c_1}^2 + m_2(l_1^2 + l_{c_2}^2 + 2l_1 l_{c_2}\cos(q_2)) \\
&+ m_3(l_1^2 + l_2^2 + l_{c_3}^2 + 2l_1 l_2 \cos(q_2) + 2l_1 l_{c_3}\cos(q_2+q_3) + 2l_2 l_{c_3}\cos(q_3)), \\
M_{12}(q) =& I_2 + I_3 + m_2(l_{c_2}^2 + 2l_1 l_{c_2}\cos(q_2)) \\
&+ m_3(l_2^2 + l_{c_3}^2 + l_1 l_2 \cos(q_2) + l_1 l_{c_3}\cos(q_2+q_3) + 2l_2 l_{c_3}\cos(q_3)), \\
M_{13}(q) =& I_3 + m_3(l_{c_3}^2 + l_1 l_{c_3}\cos(q_2+q_3) + l_2 l_{c_3}\cos(q_3)), \\
M_{21}(q) =& M_{12}(q), \\
M_{22}(q) =& I_2 + I_3 + m_2 l_{c_2}^2 + m_3(l_2^2 + l_{c_3}^2 + 2l_2 l_{c_3}\cos(q_3)),
\end{aligned}$$

$$M_{23}(q)=I_3+m_3(l_{c_3}^2+l_2l_{c_3}\cos(q_3)),$$
$$M_{31}(q)=M_{13}(q),$$
$$M_{32}(q)=M_{23}(q),$$
$$M_{33}(q)=I_3+m_3l_{c_3}^2,$$

$$C(q,\dot{q}) = \begin{bmatrix} C_{11}(q,\dot{q}) & C_{12}(q,\dot{q}) & C_{13}(q,\dot{q}) \\ C_{21}(q,\dot{q}) & C_{22}(q,\dot{q}) & C_{23}(q,\dot{q}) \\ C_{31}(q,\dot{q}) & C_{32}(q,\dot{q}) & C_{33}(q,\dot{q}) \end{bmatrix},$$

$$\begin{aligned}
C_{11}(q,\dot{q}) &= -(m_2l_1l_{c_2}\sin(q_2)+m_3l_1l_2\sin(q_2)+m_3l_1l_{c_3}\sin(q_2+q_3))\dot{q}_2 \\
&\quad -(m_3l_1l_{c_3}\sin(q_2+q_3)+m_3l_2l_{c_3}\sin(q_3))\dot{q}_3, \\
C_{12}(q,\dot{q}) &= -(m_2l_1l_{c_2}\sin(q_2)+m_3l_1l_2\sin(q_2)+m_3l_1l_{c_3}\sin(q_2+q_3))(\dot{q}_1+\dot{q}_2) \\
&\quad -(m_3l_1l_{c_3}\sin(q_2+q_3)+m_3l_2l_{c_3}\sin(q_3))\dot{q}_3, \\
C_{13}(q,\dot{q}) &= -(m_3l_1l_{c_3}\sin(q_2+q_3)+m_3l_2l_{c_3}\sin(q_3))(\dot{q}_1+\dot{q}_2+\dot{q}_3), \\
C_{2,1}(q,\dot{q}) &= (m_2l_1l_{c_2}\sin(q_2)+m_3l_1l_2\sin(q_2)+m_3l_1l_{c_3}\sin(q_2+q_3))\dot{q}_1 \\
&\quad -m_3l_2l_{c_3}\sin(q_3)\dot{q}_3, \\
C_{2,2}(q,\dot{q}) &= -m_3l_2l_{c_3}\sin(q_3)\dot{q}_3, \\
C_{2,3}(q,\dot{q}) &= -m_3l_2l_{c_3}\sin(q_3)(\dot{q}_1+\dot{q}_2+\dot{q}_3), \\
C_{3,1}(q,\dot{q}) &= (m_3l_1l_{c_3}\sin(q_2+q_3)+m_3l_2l_{c_3}\sin(q_3))\dot{q}_1+m_3l_2l_{c_3}\sin(q_3)\dot{q}_3, \\
C_{3,2}(q,\dot{q}) &= m_3l_2l_{c_3}\sin(q_3)(\dot{q}_1+\dot{q}_2), \\
C_{3,3}(q,\dot{q}) &= 0,
\end{aligned}$$

and

$$F(\dot{q}) = \begin{bmatrix} f_1\dot{q}_1 \\ f_2\dot{q}_2 \\ f_3\dot{q}_3 \end{bmatrix},$$

where m_i, l_i, l_{c_i}, I_i, and f_i are, respectively, the mass, length, center of mass, inertia, and viscous friction coefficient of link i.

The manipulator's nominal kinematic and dynamic parameters, computed from manufacturers' spec sheets and actual measurements, are presented in Table 1.2.

1.3 Control Environment for Robots (CERob)

Synthesis of control methods for robotic systems can be facilitated by utilization of software environments where one integrates algorithm design and experimental validation. Although there exist commercial programs to simulate industrial

Table 1.2 UARM kinematic and dynamic parameters

i	m_i (kg)	I_i (kg m^2)	l_i (m)	l_{c_i} (m)	f_i (kg m^2/s)
1	0.850	0.0153	0.203	0.096	0.28
2	0.850	0.0100	0.203	0.096	0.18
3	0.625	0.0100	0.203	0.077	0.10

manipulators and to automatically generate execution code, such programs tend to be expensive, not fully customizable, and unable to deal with underactuated manipulators or joint failure modes. To fill in this void we developed Control Environment for Robots (CERob), an open source MATLAB-based environment that allow us to design, test, and validate methods for manipulator fault detection and control of the resulting underactuated system. The environment includes models of single and cooperative manipulators.

The source code of this simulator is available at the Springer extra materials site (http://extras.springer.com/). After the installation procedure completes, the control environment is run by typing CERob in the MATLAB® workspace. Use of CERob assumes the user has a licensed copy of MATLAB® version 6.0 or above.

When launched, the CERob user interface appears as in Fig. 1.4. The following control environments are available for designing robust and fault tolerant controllers:

- *Underactuted Manipulator Control Environment (UMCE).* This control environment deals with single fully-actuated and underactuated manipulators. The manipulator configuration, fully-actuated or underactuated, is defined before the simulation starts, and no fault detection and isolation framework runs during simulation. Details of design procedures for the controllers available in UMCE are presented in Chaps. 2–5.
- *Fault-Tolerant Manipulator Control Environment (FTMCE).* In this control environment users can implement fault-tolerant control strategies based on Markovian controllers. The manipulator starts the movement configured as fully-actuated. When a fault is detected and isolated, the manipulator configuration changes to the corresponding fault configuration. Up to two consecutive faults can be simulated in FTMCE; details are shown in Chap. 6.
- *Cooperative Manipulator Control Environment (CMCE).* This control environment includes fault tolerant strategies and robust controllers for fully-actuated and underactuated cooperative manipulators. The underactuation configuration can be defined at the beginning of the experiment or be the result of a fault. The fault detection and isolation system is based on artificial neural networks and can be applied to free-swinging joint faults, locked joint faults, and incorrectly-measured position and velocity faults. Chapters 7–9 present the specific features of the CMCE.

The current implementation of the control environments is specific for the 3-link experimental manipulator UARM. Within these environments the user can design and test manipulator control algorithms on models that include inertial, centrifugal, Coriolis, friction and gravitational terms. At the end of each

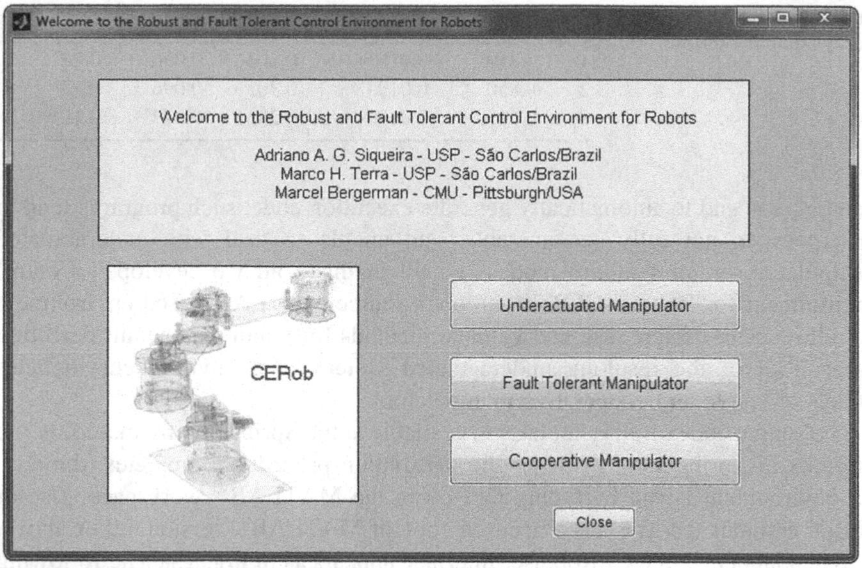

Fig. 1.4 Control Environment for Robots (CERob) initial screen

simulation, a graphic window shows the resulting joint positions, velocities and torques. Users who have an UARM system available to them may execute the exact same simulation code in the actual robot with the click of a button on the interface, and observe the resulting motion in real time.

In the remainder of this chapter we present in detail the operation of the first control environment, the Underactuated Manipulator Control Environment (UMCE, Fig. 1.5). Operation of the other two is very similar and is ommited for brevity.

The UMCE graphical user interface is divided in two areas: the frame area, which holds all the buttons used to execute commands, all messages, and a prompt for entering numeric data; and the movie axis, which displays the manipulator's motion. The frame area is sub-divided in the following areas.

- USER COMMANDS: the buttons here perform the following tasks (Fig. 1.6):
  - *Start Simulation*: initializes the simulation. Because this button is interruptible (so that one can stop the simulation at any time) it becomes invisible during the simulation. It becomes visible again when the simulation ends or when the user presses the Stop Simulation button. If the simulation fails (e.g., because of a failed matrix inversion), this button will not return to the visible state. In this case the user may simply press Stop Simulation and the button will appear again.
  - *Start UARM*: initializes an experiment with the underactuated manipulator UARM. If the interface device is not installed in the computer, an error message is displayed. The same message is also shown when the other robot-related buttons are pressed.

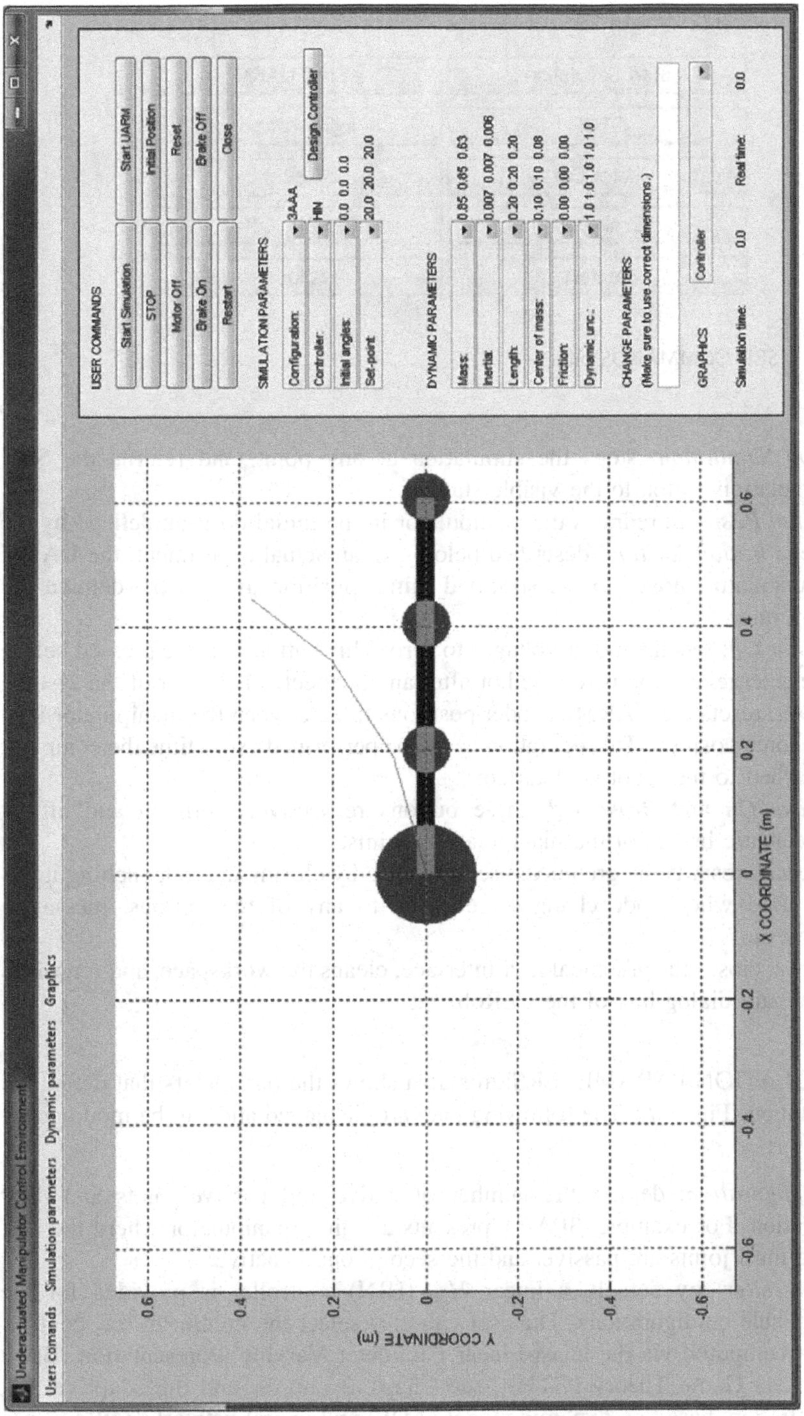

Fig. 1.5 UMCE graphical user interface

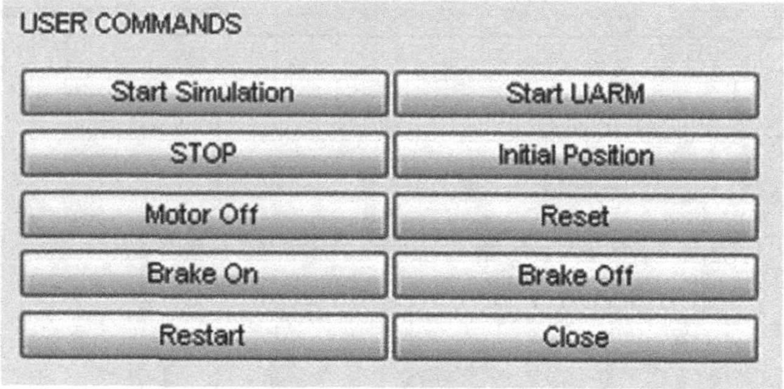

Fig. 1.6 USER COMMANDS buttons

- *Stop Simulation*: stops the simulation at any point, and returns the Start Simulation button to the visible state.
- *Initial Position*: redraws the manipulator in the initial position defined by the menu *Initial Position* (described below). In an actual experiment, the UARM manipulator moves to the specified initial position using a pre-defined PD controller.
- *Motor Off*: sets the motor voltages to zero. This button must be pressed before the emergency stop is released or after an unexpected behavior of the system.
- *Reset*: resets the UARM encoder positions to zero when the manipulator is in its home position. This operation has to be performed every time the computer attached to the robot is restarted.
- *Brake On* and *Brake Off*: these buttons respectively turn on and off all pneumatic brakes of the manipulator's joints.
- *Restart*: restarts the graphical user interface by closing and relaunching it. To be used when code changes are made to any of the buttons, messages, axes, etc.
- *Close*: closes the graphical user interface, cleans the workspace, and returns to the main dialog box of the CERob.

• SIMULATION PARAMETERS: this area shows the parameters that define the simulation (Fig. 1.7). The following data are displayed and can be modified by the user:

- *Configuration*: defines the number of active and passive joints and their location. For example, 3PAP represents a 3-link manipulator where the first and third joints are passive, and the second one is active.
- *Controller*: by default, a linear $\mathcal{H}_\infty$ (HIN) controller is provided for the available configurations. The user can also select the nonlinear $\mathcal{H}_\infty$ controllers computed via the quasi-Linear Parameter Varying representation (LPV) and via Game Theory (GTH)—see Chaps. 3 and 5—and the adaptive $\mathcal{H}_\infty$ controllers based on dynamic model (ADP) and neural networks (NET)—see

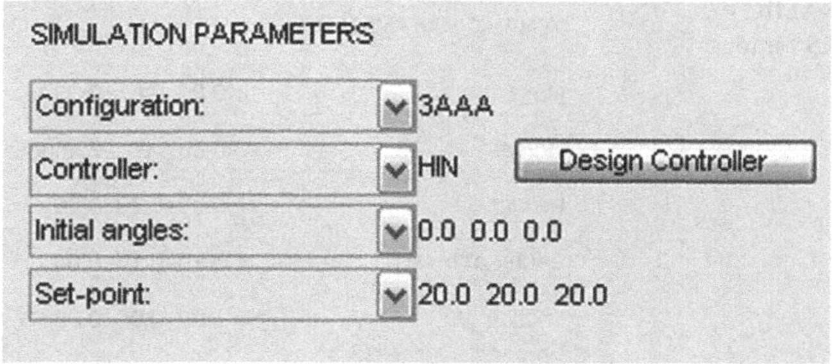

Fig. 1.7 SIMULATION PARAMETERS buttons

Chaps. 4 and 5. For the other control environments, the set of available controllers is defined in the section *Design Procedures* of the respective chapters.

– *Controller Design*: opens the controller design box, where the user can specify the control parameters and customize the available controllers.
– *Initial angles*: defines the initial angles for the manipulator. To choose the default initial conditions, click on *Default*; to choose randomly selected angles, click on *Random*. To choose your own initial angles, first enter them as a vector in the prompt provided at the message center (for example, [23.0, 34.0, −5.0]), click on *Initial angles* and click on *User defined*. Angles must be entered in degrees. If incorrect data is provided (for example, if the size of the vector entered is different from the number of joints or if the brackets are forgotten) the message "Invalid data! Default values set" will appear above the prompt and default values will be assigned to the initial conditions.
– *Set-point*: defines the set-point angles for the joints. To change the set-point, follow the procedure given above to change the initial angles.

In addition to defining initial and final joint positions using the procedures described above, the user may also directly select initial and final end-effector positions in Cartesian space by clicking anywhere on the movie axis once, and then clicking again on the desired Cartesian point. For the second click, the leftmost mouse button defines the initial position; and the center or rightmost buttons define the set-point.

In this book, joint trajectories are generated as 5th-degree polynomials connecting a joint's initial position to its set-point [2].

- DYNAMIC PARAMETERS: this area displays the parameters that define the manipulator (Fig. 1.8). The default values are those of the experimental 3-link apparatus UARM. The following data are displayed and can be modified by the user:

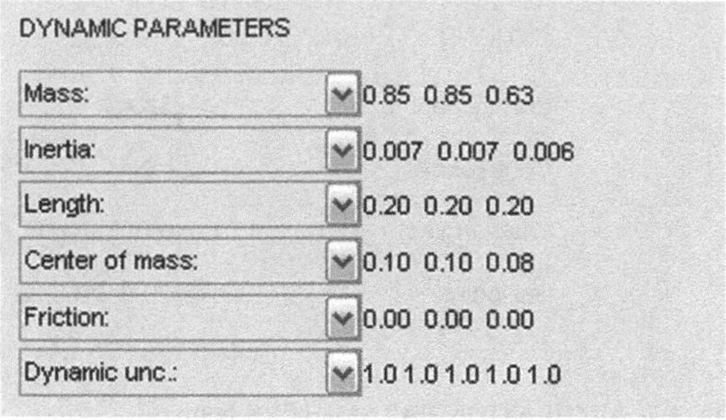

Fig. 1.8 DYNAMIC PARAMETERS buttons

- *Mass*: defines the masses of the links in kg. To change the masses, follow the procedures given above to change the initial angles.
- *Inertia*: defines the inertias of the links in kg m^2. Because we are only dealing with planar manipulators, each link's inertia is scalar. To change the inertias, follow the procedures given above to change the initial angles.
- *Length*: defines the lengths of the links in m. To change the lengths, follow the procedures given above to change the initial angles. When the user changes link lengths, the links' centers-of-mass are automatically defined to be located at the center of the link.
- *Center of mass*: defines the centers-of-mass of the links in m. To change the centers-of-mass, follow the procedures described above to change the initial angles. If the user attempts to define a center-of-mass that is not within the link's length, the message "Invalid data! Default values set" will appear above the prompt and default values will be assigned to the centers-of-mass of all links.
- *Friction*: defines the velocity-dependent friction in the joints in kg m^2/s. To change the friction, follow the procedures given above to change the initial angles.
- *Dynamic uncertainty*: defines the degree of uncertainty on the dynamic and kinematic parameters. When it is equal to 1 the dynamic model is assumed to be known perfectly. When it has a value different from 1, all kinematic and dynamic parameters are multiplied by that value and these estimated values are used by the controller. Note that the dynamic model is still computed using the nominal dynamic and kinematic parameters. This feature is provided in order to test the robustness of a particular control law with respect to parameter uncertainties.

- CHANGE PARAMETERS: provides a prompt (Fig. 1.9) for entering numeric data (as explained above). When invalid data are entered, the message "Invalid data! Default values set" will appear above the prompt and default values will be set for that particular variable.

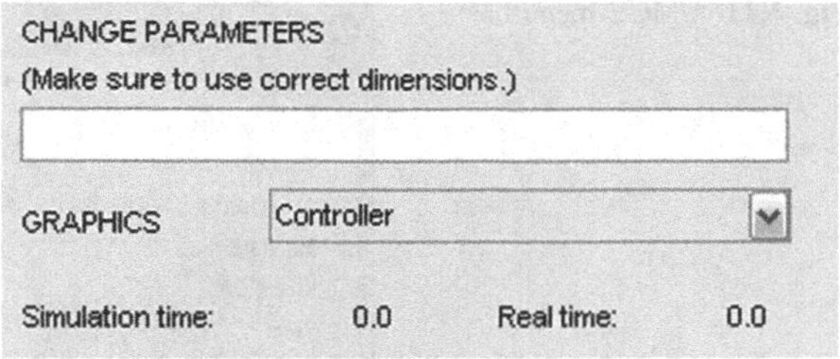

Fig. 1.9 CHANGE PARAMETERS prompt and GRAPHICS menu

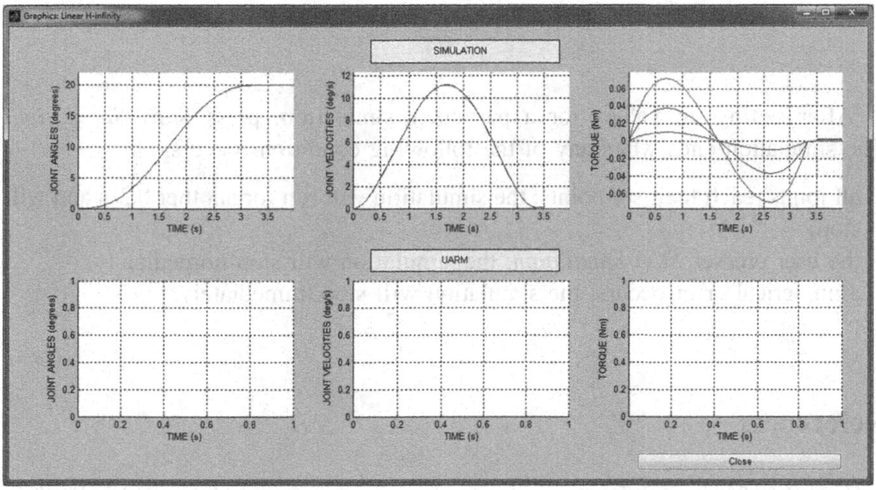

Fig. 1.10 Graphics window

- GRAPHICS: this pull-down menu displays the options available for data plotting (Fig. 1.9). Once an option is selected, a new window named "Graphics" is opened. The Graphics window displays joint positions, velocities, and torques obtained from simulation and from the experimental manipulator UARM (Fig. 1.10).

The following data are shown in the frame area and cannot be modified by the user:

- *Simulated time*: displays the simulated time as it progresses.
- *Real time*: displays the actual time of the simulation as it progresses. The wall clock time elapsed is displayed on MATLAB® command window after the simulation is over.

All the above commands are also displayed in the menu bar (Fig. 1.11).

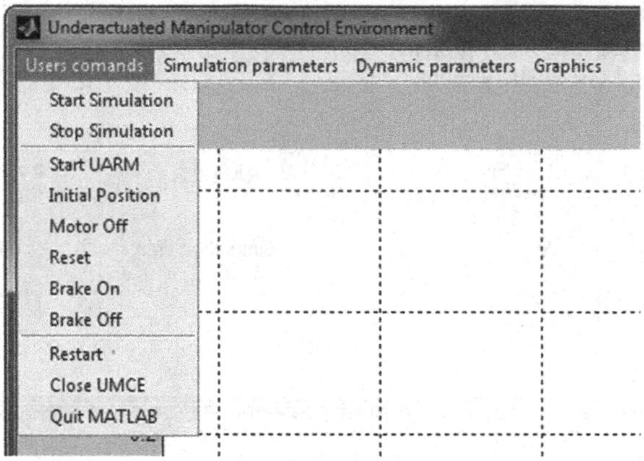

Fig. 1.11 UMCE menu bar

After setting all values for a particular simulation, press *Start Simulation*. The simulation ends when any of the following conditions occurs:

- all joints reach their set-points; the simulation will run for another 0.3 s and will stop;
- the user presses *Stop Simulation*; the simulation will stop immediately;
- a numerical error occurs; the simulation will stop immediately.

References

1. Craig JJ (1986) Introduction to robotics: mechanics and control. Addison-Wesley, Reading, MA
2. Lewis FL, Abdallah CT, Dawson DM (2004) Robot manipulator control: theory and practice. Marcel Dekker Inc, New York

Part I
Fully Actuated Robot Manipulators

Chapter 2
Linear $\mathcal{H}_\infty$ Control

2.1 Introduction

One of the most traditional techniques used for manipulator control is the computed torque method (see [6, 8, 9, 12] and references therein). It is based on the nominal dynamic model of the manipulator, and basically transforms the multivariable nonlinear plant into a set of decoupled linear systems. The computed torque method is attractive because of its simple and elegant mathematical derivation and for providing excellent control performance in the absence of modeling errors and external disturbances.

In real-world applications, however, modeling uncertainties and external disturbances degrade considerably the performance achievable by a computed torque-controlled system [11]. This can be overcome by adding an outer-loop, frequency domain-based, robust controller to the computed torque formulation. In this chapter we present one such formulation based on an $\mathcal{H}_\infty$ linear controller. We demonstrate its effectiveness by comparing it to a linear $\mathcal{H}_2$ controller. The interested reader may consult [1, 2, 4, 5, 7, 8, 14, 16–19] for more details.

This chapter is organized as follows. In Sect. 2.2 we describe the combined computed torque plus $\mathcal{H}_\infty$ linear control formulation as applied to the types of robotic manipulators presented in Chap. 1. In Sect. 2.3 we provide guidelines for synthesis of the $\mathcal{H}_\infty$ linear controller. In Sect. 2.4 we present actual results obtained when applying the formulation to the UARM.

2.2 Combined Computed Torque Plus Linear Robust Control

The fundamental concept behind computed torque control is feedback linearization of nonlinear systems. For an n-link, open chain, serial robotic manipulator, the nonlinear dynamic model is given by Eq. 1.1:

DOI: 10.1007/978-0-85729-898-0_2,

$$\tau = M(q)\ddot{q} + C(q,\dot{q})\dot{q} + F(\dot{q}) + G(q). \tag{2.1}$$

For convenience we rewrite Eq. 2.1 as:

$$\tau = M(q)\ddot{q} + b(q,\dot{q}), \tag{2.2}$$

where $b(q,\dot{q})$ is a vector composed of all non-inertial torques. In practical applications one can never know exactly the numerical values of the elements in M and b. We then redefine Eq. 2.2 so that the torque vector is now a function of the estimate of the inertia matrix and non-inertial torques:

$$\tau = \hat{M}(q)\tau' + \hat{b}(q,\dot{q}). \tag{2.3}$$

Assuming that the manipulator is programmed to follow a desired trajectory defined a priori, we can compute τ' based on the classical proportional-derivative (PD) controller:

$$\tau' = K_p(q^d - q) + K_d(\dot{q}^d - \dot{q}) + \ddot{q}^d, \tag{2.4}$$

where q^d, $\dot{q}^d$ and $\ddot{q}^d$ are the desired trajectory's position, velocity and acceleration, respectively; $K_p \in \mathbb{R}^{n\times n}$ and $K_d \in \mathbb{R}^{n\times n}$ are the proportional and derivative gains (in this chapter designed as diagonal matrices). Combining Eqs. 2.2–2.4 we obtain:

$$\ddot{e} + K_d\dot{e} + K_p e = \hat{M}^{-1}(q)[(M(q) - \hat{M}(q))\ddot{q} + (b(q,\dot{q}) - \hat{b}(q,\dot{q}))]. \tag{2.5}$$

where e is the position error $e = q^d - q$. In addition to modeling uncertainties, real-world robots are subject to internal and external disturbances such as friction, torque discretization, and load variations in the end-effector, among others. We lump all these effects in a disturbance torque vector w and add them to Eq. 2.5 as:

$$\ddot{e} + K_d\dot{e} + K_p e = \hat{M}(q)^{-1}[(M(q) - \hat{M}(q))\ddot{q} + (b(q,\dot{q}) - \hat{b}(q,\dot{q})) + w]. \tag{2.6}$$

Note in Eq. 2.6 that if we had perfect knowledge of the manipulator parameters ($M(q) = \hat{M}(q)$ and $b(q,\dot{q}) = \hat{b}(q,\dot{q})$) and if there did not exist external disturbances ($w = 0$), the right side would vanish and the computed torque method would provide perfect trajectory tracking. The state and output equations for this ideal system, controlled through an input signal $u(t)$, are given by:

$$\dot{\tilde{x}} = A_p\tilde{x} + B_p u, \tag{2.7}$$

$$y = C_p\tilde{x} + D_p u, \tag{2.8}$$

where $\tilde{x} \in \mathbb{R}^m$ is the state of the manipulator defined as

$$\tilde{x} = \begin{bmatrix} e \\ \dot{e} \end{bmatrix} = \begin{bmatrix} q^d - q \\ \dot{q}^d - \dot{q} \end{bmatrix}, \tag{2.9}$$

and the matrices $A_p \in \mathbb{R}^{m\times m}$, $B_p \in \mathbb{R}^{m\times n}$, $C_p \in \mathbb{R}^{n\times m}$ and $D_p \in \mathbb{R}^{n\times n}$ as

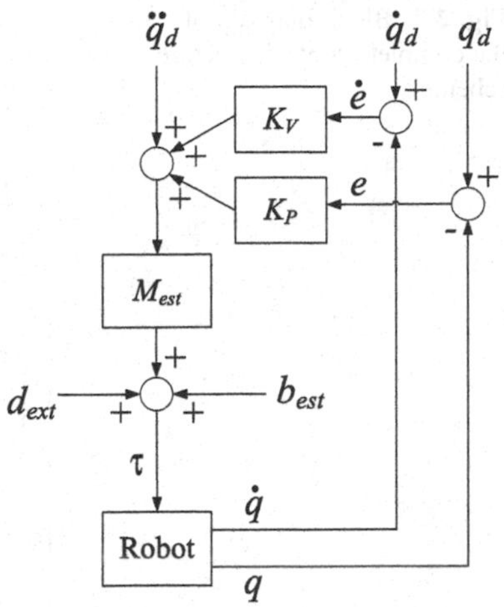

Fig. 2.1 Block diagram of the computed torque controller applied to an ideal, disturbance-free manipulator

$$A_p = \begin{bmatrix} 0 & I \\ -K_p & -K_d \end{bmatrix}, \quad B_p = \begin{bmatrix} 0 \\ I \end{bmatrix}, \quad C_p = [I \quad 0],$$

and $D_p = 0$. Figure 2.1 presents a block diagram of this idealized control system.

To deal with modeling uncertainties and external disturbances, the computed torque controller can be combined with a frequency domain-based outer-loop controller, $K(s)$, as illustrated in Fig. 2.2, where K^* represents its time-domain realization.

The weighting function $W_e(s)$ is used to shape the system performance in the frequency domain. The weighting function $W_\Delta(s)$ is used to shape the multiplicative unstructured uncertainties in the input of the system. Details of structured and unstructured uncertainties concepts can be found in [17]. The portion of Fig. 2.2 delimited by the dotted line is the augmented plant $P(s)$, used to design linear controllers (e.g., $\mathcal{H}_2$, $\mathcal{H}_\infty$, and $\mathcal{H}_2/\mathcal{H}_\infty$). Note that the plant $P(s)$ and the weighting functions $W_e(s)$ and $W_\Delta(s)$ are also represented in Fig. 2.2 by their time-domain realizations P^*, W_e^*, and $W_\Delta*$, respectively.

Here we present a design procedure based on the $\mathcal{H}_\infty$ formulation. (An $\mathcal{H}_2$ controller can be easily derived from the $\mathcal{H}_\infty$ one, as we will show in the sequel.) The design of the robust controller is performed in two steps. First, the computed torque method is used to pre-compensate the dynamics of the nominal system. Then, the $\mathcal{H}_\infty$ controller is used to post-compensate the residual error which is not completely removed by the computed torque method. As we will show, the combined controller is able to perform robust tracking control.

We start by finding a state-space realization of the augmented plant $P(s)$ through the definition of K_p, K_d, $W_e(s)$, and $W_\Delta(s)$. The performance objectives $W_e(s)$ and $W_\Delta(s)$ are respectively related to the frequency response of the sensitivity function:

$$S(s) = (I + P(s)K(s))^{-1},$$

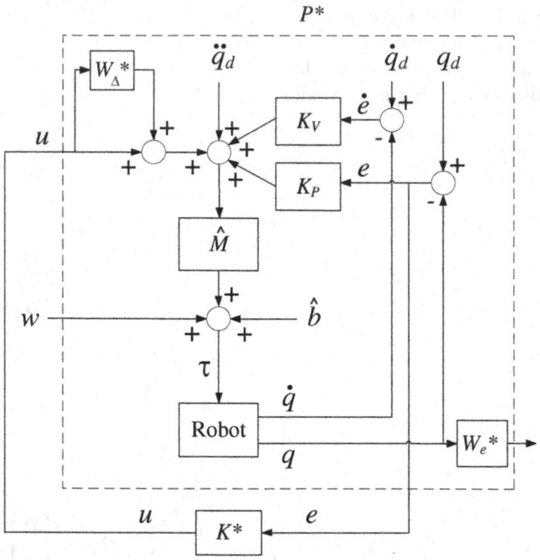

Fig. 2.2 Block diagram of the complete control structure scheme

where $K(s)$ is the robust controller, and of the complementary sensitivity function

$$T(s) = I - S(s).$$

To define $W_e(s)$, we select a bandwidth ω_b, a maximum peak M_s, and a small $\epsilon > 0$. With these specifications in hand, the following performance-shaping, diagonal weighting matrix can be determined:

$$W_e(s) = \begin{bmatrix} W_{e,1}(s) & 0 & \dots & 0 \\ 0 & W_{e,2}(s) & \dots & 0 \\ \vdots & \vdots & \ddots & \vdots \\ 0 & 0 & \dots & W_{e,n}(s) \end{bmatrix}, \tag{2.10}$$

$$W_{e,i}(s) = \frac{s + \omega_b}{M_s(s + \omega_b \epsilon)},$$

where $i = 1, \dots, n$. To define $W_\Delta(s)$, we select the maximum gain M_u of $K(s)S(s)$, the controller bandwidth ω_{bc} and a small $\epsilon_1 > 0$ such that:

$$W_\Delta(s) = \begin{bmatrix} W_{\Delta,1}(s) & 0 & \dots & 0 \\ 0 & W_{\Delta,2}(s) & \dots & 0 \\ \vdots & \vdots & \ddots & \vdots \\ 0 & 0 & \dots & W_{\Delta,n}(s) \end{bmatrix}, \tag{2.11}$$

$$W_{\Delta,i}(s) = \frac{s + \omega_{bc}}{M_u(\epsilon_1 s + \omega_{bc})}.$$

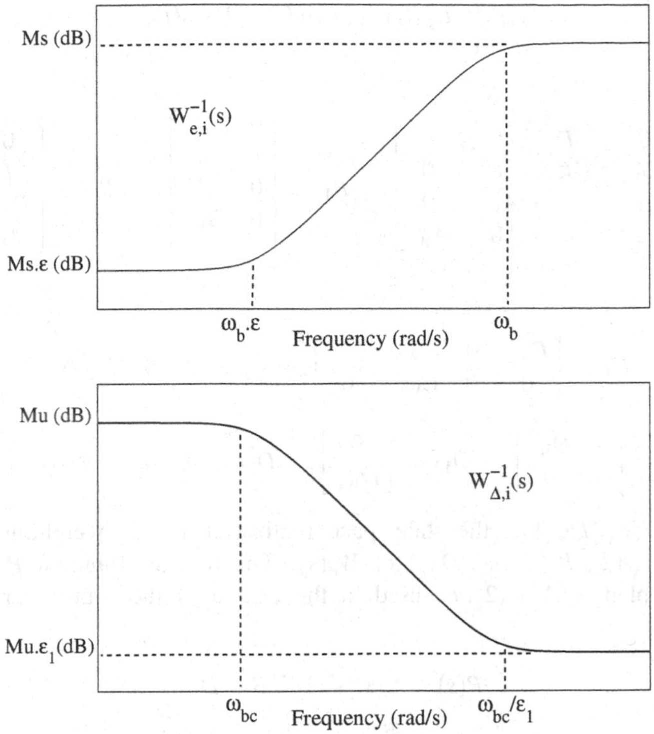

Fig. 2.3 Frequency responses of $W_e^{-1}(s)$ and $W_\Delta^{-1}(s)$

Figure 2.3 shows the frequency responses of $W_e^{-1}(s)$ and $W_\Delta^{-1}(s)$. In the $\mathcal{H}_\infty$ design procedure, they are selected in order to guarantee:

$$||S||_\infty \leq ||W_e^{-1}||_\infty \tag{2.12}$$

and

$$||T||_\infty \leq ||W_\Delta^{-1}||_\infty. \tag{2.13}$$

The $\mathcal{H}_\infty$ norm of a given transfer function $G(s)$ is defined as:

$$||G||_\infty = \sup_{\omega \in \mathbb{R}} \bar{\sigma}\{G(j\omega)\}, \tag{2.14}$$

where $\sup \bar{\sigma}\{.\}$ denotes the supremum of the maximum singular value.

As we are interested in the state-space form of the augmented system $P(s)$, we redefine the control system in the time domain as:

$$\dot{x}(t) = Ax(t) + B_1 w(t) + B_2 u(t), \tag{2.15}$$

$$z(t) = C_1 x(t) + D_{11} w(t) + D_{12} u(t), \tag{2.16}$$

$$y(t) = C_2 x(t) + D_{21} w(t) + D_{22} u(t), \tag{2.17}$$

where

$$A = \begin{bmatrix} 0 & I & 0 & 0 \\ -K_p & -K_d & 0 & 0 \\ 0 & 0 & A_{W_\Delta} & 0 \\ B_{W_e} & 0 & 0 & A_{W_e} \end{bmatrix}, \quad B_1 = \begin{bmatrix} 0 & 0 \\ I & 0 \\ 0 & 0 \\ 0 & B_{W_e} \end{bmatrix}, \quad B_2 = \begin{bmatrix} 0 \\ I \\ B_{W_\Delta} \\ 0 \end{bmatrix}, \tag{2.18}$$

$$C_1 = \begin{bmatrix} D_{W_e} & 0 & 0 & C_{W_e} \\ 0 & 0 & C_{W_\Delta} & 0 \end{bmatrix}, \quad C_2 = [I \ \ 0 \ \ 0 \ \ 0],$$

$$D_{11} = \begin{bmatrix} 0 & D_{W_e} \\ 0 & 0 \end{bmatrix}, \quad D_{12} = \begin{bmatrix} 0 \\ D_{W_\Delta} \end{bmatrix}, \quad D_{21} = [0 \ \ I], \quad D_{22} = [0],$$

$(A_{W_\Delta}, B_{W_\Delta}, C_{W_\Delta}, D_{W_\Delta})$ is the state-space realization of the weighting function $W_\Delta(s)$, and $(A_{W_e}, B_{W_e}, C_{W_e}, D_{W_e})$ of $W_e(s)$. The transfer function $P(s)$ of the augmented plant (2.15)–(2.17), used in the design of the robust controller, is given by:

$$P(s) = C(sI - A)^{-1} B + D, \tag{2.19}$$

where

$$C = \begin{bmatrix} D_{W_e} & 0 & 0 & C_{W_e} \\ 0 & 0 & C_{W_\Delta} & 0 \\ I & 0 & 0 & 0 \end{bmatrix}, \quad B = \begin{bmatrix} 0 & 0 & 0 \\ I & 0 & I \\ 0 & 0 & B_{W_\Delta} \\ 0 & B_{W_e} & 0 \end{bmatrix},$$

$$D = \begin{bmatrix} 0 & D_{W_e} & 0 \\ 0 & 0 & D_{W_\Delta} \\ 0 & I & 0 \end{bmatrix},$$

and the matrix A is given in (2.18).

2.3 Linear $\mathcal{H}_\infty$ Controllers

In this section we present the basics of linear $\mathcal{H}_\infty$ control design. We refer the reader to the vast literature on the subject for more details (e.g., [14, 16]). The system is represented by the block diagram in Fig. 2.4, which shows the plant $P(s)$ and the controller $K(s)$. The plant has two sets of input signals, the internal input u and the external input w, and two sets of output signals, the measured signal y and the regulated output z.

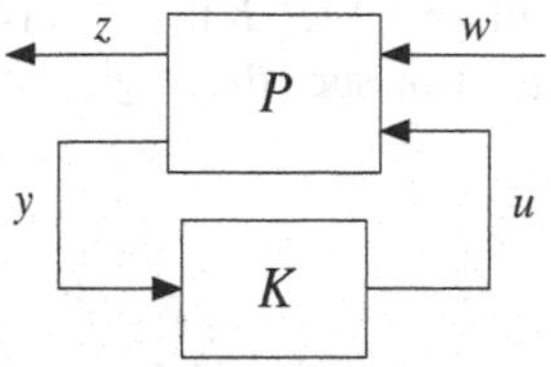

Fig. 2.4 Block diagram for $\mathcal{H}_\infty$ control systems

The objective of an $\mathcal{H}_\infty$ controller is to guarantee that the $\mathcal{H}_\infty$ norm of a multivariable transfer function $T_{zw}(s)$ is limited by a level of attenuation γ, $\|T_{zw}(s)\|_\infty < \gamma$. The parameter γ indicates the level of robustness of the control system, or how much the input disturbances are attenuated in the output of the system. The following assumptions are required to design a simplified version of the $\mathcal{H}_\infty$ controller, based on the system (2.15)–(2.17):

(A1) (A, B_2) is stabilizable and (C_2, A) is detectable;
(A2) $D_{11} = 0$ and $D_{22} = 0$;
(A3) $D_{12}^T C_1 = 0$ and $B_1 D_{21}^T = 0$;
(A4) $D_{12} = \begin{bmatrix} 0 \\ I \end{bmatrix}$ and $D_{21} = \begin{bmatrix} 0 & I \end{bmatrix}$;
(A5) $\begin{bmatrix} A - j\omega I & B_2 \\ C_1 & D_{12} \end{bmatrix}$ has full column rank for all $\omega \in \mathbb{R}$;
(A6) $\begin{bmatrix} A - j\omega I & B_1 \\ C_2 & D_{21} \end{bmatrix}$ has full row rank for all $\omega \in \mathbb{R}$.

To synthesize the $\mathcal{H}_\infty$ controller we need the X_∞ and Y_∞ that solve the following two algebraic Riccati equations associated with the state feedback control and the state estimate of the robot:

$$X_\infty(A - B_2 D_{12}^T C_1) + (A - B_2 D_{12}^T C_1)^T X_\infty + X_\infty(\gamma^{-2} B_1 B_1^T - B_2 B_2^T) X_\infty + \hat{C}_1^T \hat{C}_1 = 0 \qquad (2.20)$$

and

$$(A - B_1 D_{21}^T C_2) Y_\infty + Y_\infty (A - B_1 D_{21}^T C_2)^T + Y_\infty(\gamma^{-2} C_1 C_1^T - C_2 C_2^T) Y_\infty + \hat{B}_1 \hat{B}_1^T = 0, \qquad (2.21)$$

where $\hat{C}_1 = (I - D_{12} D_{12}^T) C_1$ and $\hat{B}_1 = B_1 (I - D_{21}^T D_{21})$.

A stabilizing solution for this controller can be found if the matrices X_∞ and Y_∞ are positive semi-definite and the spectral radius of $X_\infty Y_\infty$ satisfies $\rho(X_\infty Y_\infty) \leq \gamma^2$. The design problem consists of finding minimum γ that obeys this inequality, thus yielding the "best" robustness. The family of all stabilizing controllers K_∞ that satisfy $||\mathcal{F}(P,K)||_\infty \leq \gamma$ is given by $K_\infty = \mathcal{F}(J,Q)$ where Q is any stable transfer function such that $||Q||_\infty < \gamma$, $\mathcal{F}(...)$ represents a linear fractional transformation, and

$$J = \begin{bmatrix} J_{11} & J_{12} \\ J_{21} & J_{22} \end{bmatrix}, \tag{2.22}$$

where

$$J_{11} = A + B_2 F_\infty + \gamma^{-2} B_1 B_1^T X_\infty + Z_\infty H_\infty (C_2 + \gamma^{-2} D_{21} B_1^T X_\infty),$$

$$J_{12} = \left[-Z_\infty H_\infty - Z_\infty (B_2 + \gamma^{-2} Y_\infty C_1^T D_{12} \right],$$

$$J_{21} = \begin{bmatrix} F\infty \\ -(C_2 + \gamma^{-2} D_{12} B_1^T X_\infty) \end{bmatrix}, \quad J_{22} = \begin{bmatrix} 0 & I \\ I & 0 \end{bmatrix}$$

$$F_\infty = -(B_2^T X_\infty + D_{12}^T C_1), \quad H_\infty = -(Y_\infty C_2^T + B_1 D_{21}^T),$$

$$Z_\infty = (I - \gamma^{-2} Y_\infty X_\infty)^{-1}.$$

Note that the matrices in (2.19) do not always satisfy the assumptions aforementioned. In [14], and references therein, the authors provide a procedure to guarantee that these assumptions hold. This procedure is based on strict system equivalence transformations. The MATLAB® function *hinfsyn*, which is used to design the $\mathcal{H}_\infty$ controller, incorporates the algorithm used in that procedure. We omit the details here for brevity. In the next section we show examples of how to use the function *hinfsyn*.

According to [7], $\mathcal{H}_2$ controllers are a special case of $\mathcal{H}_\infty$ controllers. By setting the attenuation level γ to ∞ in Eqs. 2.20 and 2.21, the resulting Ricatti equations become the ones of classical $\mathcal{H}_2$ controllers. We use this result in the following Examples section.

2.4 Examples

In this section we present practical examples that illustrate the application of the combined computed torque plus linear $\mathcal{H}_\infty$ controller to the UARM robotic manipulator, including the MATLAB® code used to design the robust controller.

2.4.1 Design Procedures

The controller design procedure is automated within the CERob simulator. After launching the software, select the *Controller Design* option in the *Controller* menu. The design dialog box is shown in Fig. 2.5.

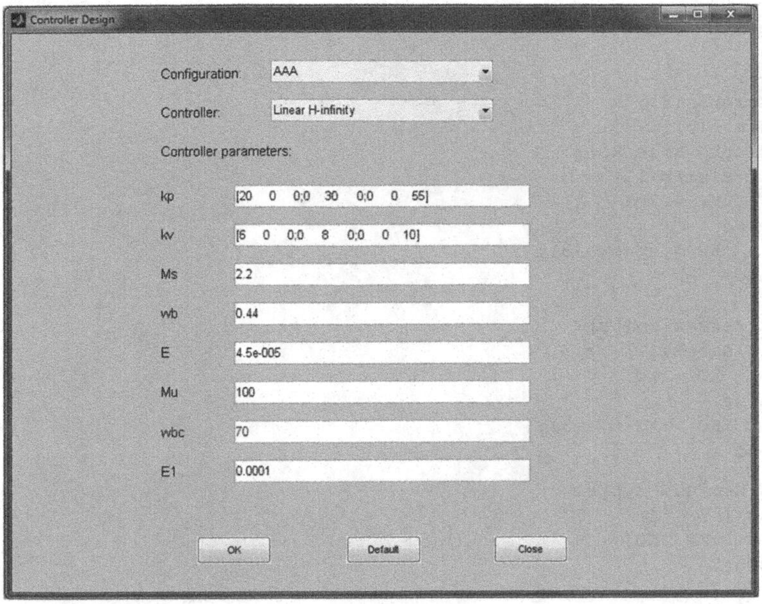

Fig. 2.5 Controller Design box for linear $\mathcal{H}_\infty$ controller

As discussed previously in this chapter, the design procedure can be summarized as:

1. Select the gains K_p and K_d and compute the nominal plant (A_p, B_p, C_p, and D_p) given by Eq. 2.7;
2. Select M_s, ω_b and ε in (2.10) and M_u, ω_{bc} and ε_1 in (2.11);
3. Compute the augmented plant (2.15)–(2.17) and $\mathcal{H}_\infty$ controller;
4. Compute the sensitivity function $S(s)$ and the complementary sensitivity function $T(s)$. Plot the graph of $S(s)$ versus $W_e^{-1}(s)$ and $T(s)$ versus $W_\Delta^{-1}(s)$;
5. Check that conditions (2.12) and (2.13), respectively, are satisfied; if not, return to step 1.

These steps are performed by the *CERob* control environment using the MATLAB® Robust Control Toolbox. The commands *sysic* and *hinfsyn* are used to generate the augmented plant and to design the $\mathcal{H}_\infty$ controller, respectively. Details of the algorithms used in these functions can be found in [1, 5]. The following box shows the code used to design the controller (available in the directory `CERob\UnderactuatedSimulator`).

File: HIN_3AAA.m

```
...

Z=zeros(3);
I=eye(3);
sizekp=size(kp);
if sizekp(1) == 3
    KP = kp;
else
    KP = kp*eye(3);
end

sizekv=size(kv);
if sizekv(1) == 3
    KD = kd;
else
    KD = kd*eye(3);
end

% Nominal system
Ag=[ Z   I
    -KP -KD];
Bg=[Z
    I];
Cg=100*[I Z];
Dg=Z;
G=pck(Ag,Bg,Cg,Dg);

% Weighting definition
nwe=(1/Ms)[1 wb];
dwe=[1 wb*E];
We=tf(nwe,dwe)
We1=nd2sys(nwe,dwe);
We2=nd2sys(nwe,dwe);
We3=nd2sys(nwe,dwe);
We=daug(We1,We2,We3);

nwu=(1/Mu)[1 wbc];
dwu=[E1 wbc];
Wu=tf(nwu,dwu)
Wu1=nd2sys(nwu,dwu);
Wu2=nd2sys(nwu,dwu);
Wu3=nd2sys(nwu,dwu);
Wu=daug(Wu1,Wu2,Wu3);

% Augmented plant
systemnames='G Wu We';
inputvar='[pert(3); dist(3); control(3)]';
outputvar='[Wu ; We; G+dist]';
input_to_G='[control+pert]';
input_to_Wu='[control]';
input_to_We='[G+dist]';
sysoutname='P';
cleanupsysic='yes';
sysic
```

```
% H-infinity synthesis
gmin=0.1;
gmax=150;
K1 = hinfsyn(P,3,3,gmin,gmax,0.05);
[Ak,Bk,Ck,Dk]=unpck(K1);
K = ss(Ak,Bk,Ck,Dk);

% Graphics
[Agk,Bgk,Cgk,Dgk] = series(Ak,Bk,Ck,Dk,Ag,Bg,Cg,Dg);
Lo = ss(Agk,Bgk,Cgk,Dgk);

% Sensitivity function
So = inv(eye(size(Lo))+Lo);

We = tf(nwe,dwe);
WE = eye(3)*We;
Wu = tf(nwu,dwu);
WU = eye(3)*Wu;
figure
wsigma = logspace(log10(wb*E)-2,log10(wb)+2,100);
SV1 = sigma(So,wsigma);
minSV = min(min(20*log10(SV1)));
maxSV = max(max(20*log10(SV1)))
semilogx(wsigma,20*log10(SV1(1,:)),'k');
hold on
SV = sigma(inv(WE),wsigma);
semilogx(wsigma,20*log10(SV),'k--');
semilogx(wsigma,20*log10(SV1(2:3,:)),'k');

...

% Complementary sensitivity function
To = series(Lo,So);

figure
wsigma = logspace(log10(wbc)-2,log10(wbc/E1)+2,100);
SV1 = sigma(To,wsigma);
minSV = min(min(20*log10(SV1)));
maxSV = max(max(20*log10(SV1)))
semilogx(wsigma,20*log10(SV1(1,:)),'k');
hold on
SV = sigma(inv(Wu),wsigma);
semilogx(wsigma,20*log10(SV),'k--');
semilogx(wsigma,20*log10(SV1(2:3,:)),'k');
...
```

Figure 2.6 presents the resulting sensitivity function $S(s)$ and the complementary sensitivity function $T(s)$, computed using the default parameters shown in Fig. 2.5. Note that both $S(s)$ and $T(s)$ are strictly lower than the weighting functions $W_e^{-1}(s)$ and $W_\Delta^{-1}(s)$, respectively, for all frequencies and therefore satisfy Eqs. 2.12 and 2.13. The $\mathcal{H}_2$ controller is designed based on the same code considering $\gamma \rightarrow \infty$.

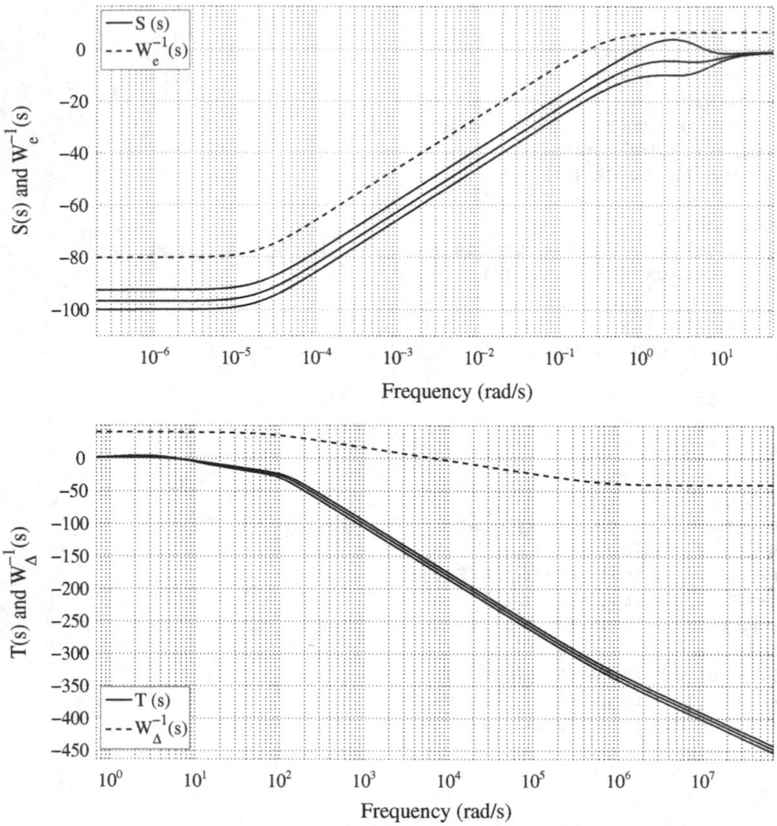

Fig. 2.6 Sensibility function $S(s)$ versus $W_e^{-1}(s)$ and complementary sensibility function $T(s)$ versus $W_\Delta^{-1}(s)$

2.4.2 Experimental Results

The combined computed torque plus $\mathcal{H}_\infty$ or $\mathcal{H}_2$ controllers were applied to the UARM on a trajectory with starting position $q(0) = [0° \ 0° \ 0°]^T$ and desired position $q(T) = [-20° \ 30° \ -30°]^T$ with $T = [4.0 \quad 4.0 \quad 4.0]$s. To test the controllers' disturbance rejection properties we add, to the torques computed by the robust controllers, exponentially attenuated sinusoidal torque disturbances of the form:

$$\tau_{d_i} = A_i e^{\frac{-(t-t_{f_i})^2}{2\sigma_i^2}} \sin(\omega_i t),$$

where A_i is the maximum disturbance amplitude for joint i, t_{f_i} and σ_i are respectively the mean and standard deviation of the attenuation function, and ω_i is the frequency of the sinusoid. Figure 2.7 presents the torque disturbances for

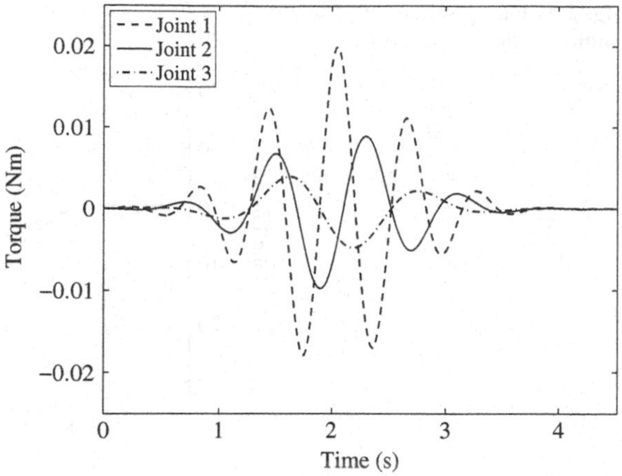

Fig. 2.7 Torque disturbances

$A = [0.02 \quad -0.01 \quad 0.005]$ N, $\omega = [10 \quad 7.5 \quad 5]$ rad/s, $t_{f_i} = 2$ s, and $\sigma_i = 0.6$, for $i = 1, \ldots, 3$. These values were chosen so that the maximum values of the torque disturbances represent at least 20% of the maximum applied torques during experiments performed without disturbances.

The parameters used to design the linear $\mathcal{H}_2$ and $\mathcal{H}_\infty$ controllers are presented in Tables 2.1 and 2.2. The experimental results obtained when the manipulator was not subject to torque disturbances are shown in Figs. 2.8, 2.9, 2.10 and 2.11. Note that the joint positions track the desired trajectories and reach the set-points within the specified time.

The experimental results for the case where external disturbances are applied are shown in Figs. 2.12, 2.13, 2.14 and 2.15. One can clearly note the decrease in performance, which is anticipated given the high disturbance values compared to the nominal torques. Still, the joint positions reach the set-points once the disturbances are attenuated.

Table 2.1 K_p and K_d gains for the $\mathcal{H}_2$ and $\mathcal{H}_\infty$ controllers

Controller	K_p	K_d
Linear $\mathcal{H}_2$ control	$\begin{bmatrix} 20 & 0 & 0 \\ 0 & 30 & 0 \\ 0 & 0 & 50 \end{bmatrix}$	$\begin{bmatrix} 6 & 0 & 0 \\ 0 & 8 & 0 \\ 0 & 0 & 10 \end{bmatrix}$
Linear $\mathcal{H}_\infty$ control	$\begin{bmatrix} 20 & 0 & 0 \\ 0 & 30 & 0 \\ 0 & 0 & 55 \end{bmatrix}$	$\begin{bmatrix} 6 & 0 & 0 \\ 0 & 8 & 0 \\ 0 & 0 & 10.5 \end{bmatrix}$

Table 2.2 Performance parameters for the $\mathcal{H}_2$ and $\mathcal{H}_\infty$ controllers

Controller	M_s	ω_b (rad/s)	ε	M_u	ω_{bc} (rad/s)	ε_1
$\mathcal{H}_2$ and $\mathcal{H}_\infty$ controls	2.2	0.44	10^{-5}	100	70	10^{-4}

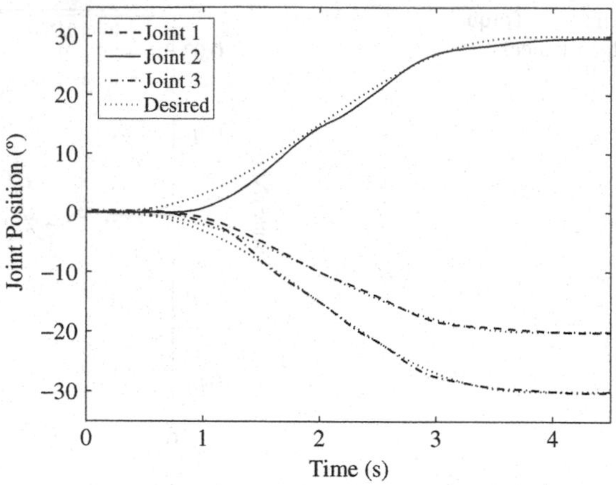

Fig. 2.8 Joint positions, $\mathcal{H}_2$ control, without disturbances

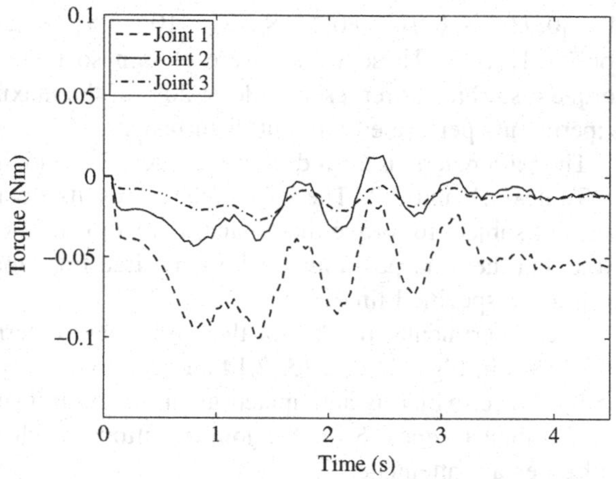

Fig. 2.9 Applied torques, $\mathcal{H}_2$ control, without disturbances

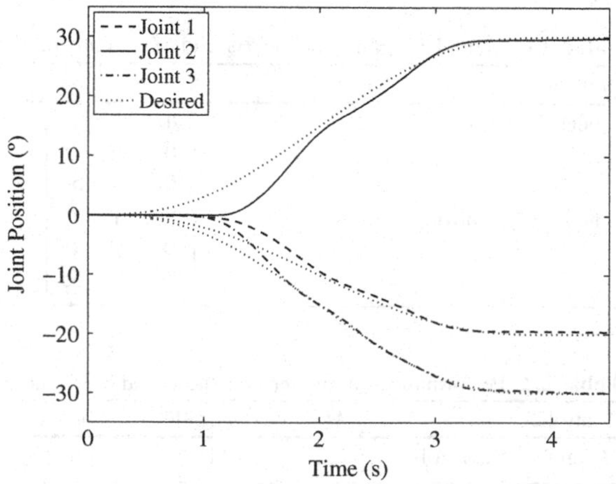

Fig. 2.10 Joint positions, $\mathcal{H}_\infty$ control, without disturbances

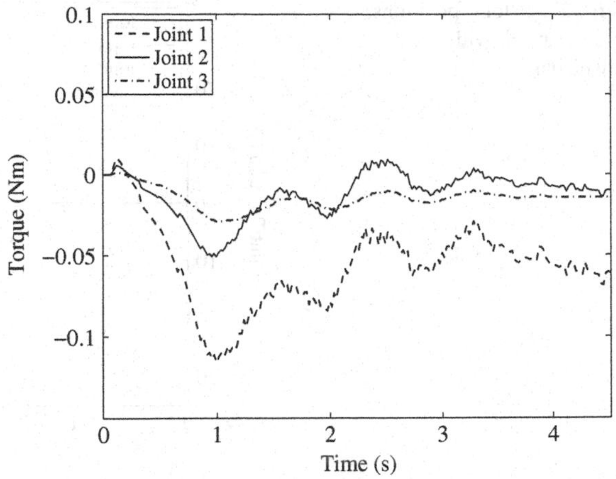

Fig. 2.11 Applied torques, $\mathcal{H}_\infty$ control, without disturbances

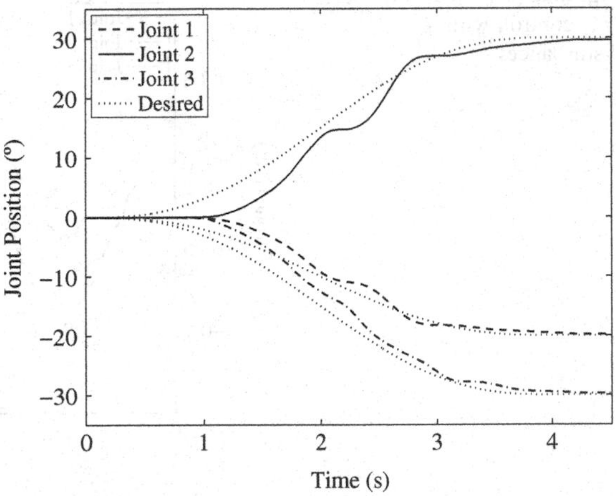

Fig. 2.12 Joint positions, $\mathcal{H}_2$ control, with disturbances

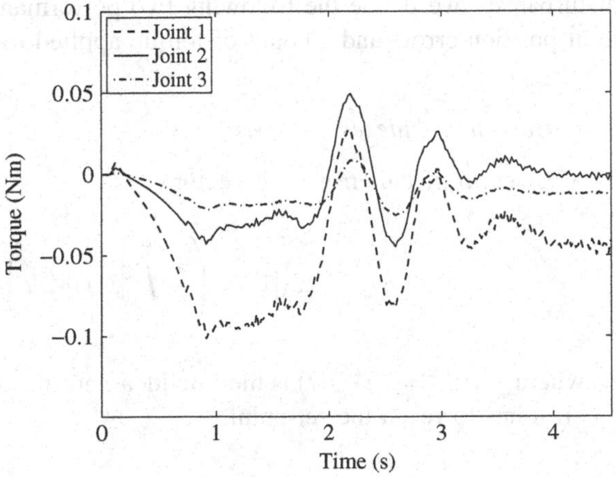

Fig. 2.13 Applied torques, $\mathcal{H}_2$ control, with disturbances

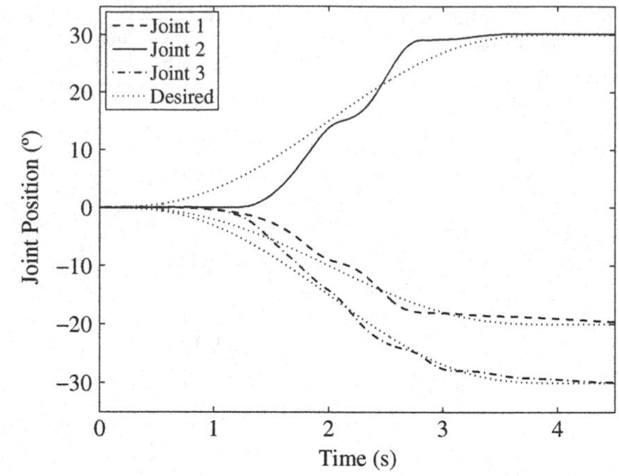

Fig. 2.14 Joint positions, $\mathcal{H}_\infty$ control, with disturbances

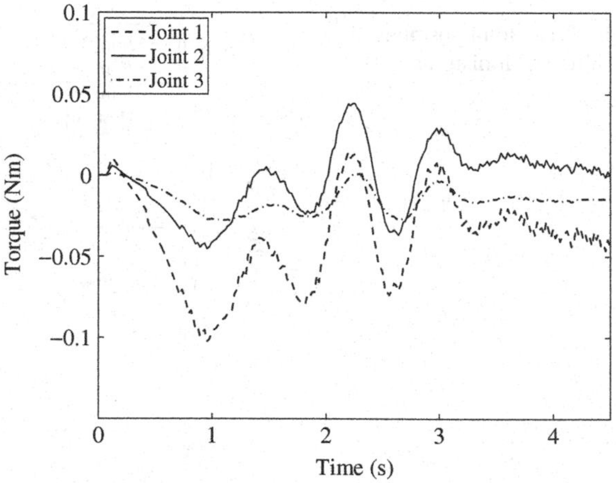

Fig. 2.15 Applied torques, $\mathcal{H}_\infty$ control, with disturbances

To compare the performance of $\mathcal{H}_\infty$ and $\mathcal{H}_2$ controllers with and without disturbances, we define the following two performance indexes that measure the joint position errors and amount of torque applied to each joint:

Performance Indexes

- $\mathcal{L}_2$ norm of the error state vector

$$\mathcal{L}_2[\tilde{x}] = \left(\frac{1}{t_r} \int_0^{t_r} \|\tilde{x}(t)\|_2^2 dt \right)^{\frac{1}{2}}, \tag{2.23}$$

where $\|\tilde{x}(t)\|^2 = \tilde{x}(t)^T \tilde{x}(t)$ is the Euclidean norm and t_r is the time it takes for all joints to reach the set-point.

- Sum of the applied torques

$$E[\tau] = \sum_{i=0}^{n} \left(\int_{0}^{t_r} |\tau_i(t)| dt \right). \tag{2.24}$$

Table 2.3 Comparison of the performance of the $\mathcal{H}_\infty$ and $\mathcal{H}_2$ controllers with and without torque disturbances

Controller	$\mathcal{L}_2[\tilde{x}]$	$E[\tau]$ (Nms)
$\mathcal{H}_2$ control, without disturbance	0.0168	0.3509
$\mathcal{H}_\infty$ control, without disturbance	0.0195	0.3429
$\mathcal{H}_2$ control, with disturbance	0.0212	0.3776
$\mathcal{H}_\infty$ control, with disturbance	0.0208	0.3506

The $\mathcal{L}_2$ norm measures the joint position errors and is widely used to compare the performance of controllers (see, for example, [3, 10, 13, 15]). The sum of applied torques index is an indirect measure of the energy consumption of the robot.

We repeated the experiments described above five times, and computed the average of each performance index; the results are shown in Table 2.3. The experimental results shown in Figs. 2.8, 2.9, 2.10, 2.11, 2.12, 2.13, 2.14 and 2.15 correspond to the samples that are closest to the mean values of $\mathcal{L}_2(\tilde{x})$ and $E[\tau]$ shown in Table 2.3.

Note that when the manipulator is subject to disturbances $\mathcal{L}_2[x]$ norm of the linear $\mathcal{H}_2$ controller increases in 26%, whereas the increase of the same index for the $\mathcal{H}_\infty$ controller was only 7%. The increase in sum of applied torques was approximately 8% and 2% for $\mathcal{H}_2$ and $\mathcal{H}_\infty$ controllers, respectively. In other words, the $\mathcal{H}_\infty$ controller provides greater robustness to external disturbances. This example illustrates the main motivation of this book, that is, to develop linear and nonlinear robust controllers based on $\mathcal{H}_\infty$ approaches for robotic manipulators.

References

1. Balas GJ, Doyle JC, Glover K, Packard A, Smith R (2001) μ-Analysis and synthesis toolbox. The MathWorks, Inc., Natick
2. Beaven RW, Wright MT, Seaward DR (1996) Weighting function selection in the $\mathcal{H}_\infty$ design process. Control Eng Pract 4(5):625–633
3. Berghuis H, Roebbers H, Nijmeijer H (1995) Experimental comparison of parameter estimation methods in adaptive robot control. Automatica 31(9):1275–1285

4. Bernstein DS, Haddad WM (1989) LQG control with an $\mathcal{H}_\infty$ performance bound: a Riccati equation approach. IEEE Trans Autom Control 34(3):293–305
5. Chiang RY, Safonov MG (1992) Robust control toolbox. The MathWorks, Inc., Natick
6. Craig JJ (1986) Introduction to robotics: mechanics and control. Addison-Wesley, Reading, MA
7. Doyle JC, Glover K, Khargonekar PP, Francis BA (1989) State space solutions to standard $\mathcal{H}_2$ and $\mathcal{H}_\infty$ problems. IEEE Trans Autom Control 34(3):831-847
8. Gilbert EG, Ha IJ (1984) An approach to nonlinear feedback control with applications to robotics. IEEE Trans Syst Man Cybern 14(6):879–884
9. Hunt LR, Sun R, Meyer G (1983) Global transformations of nonlinear systems. IEEE Trans Autom Control 28(1):24–31
10. Jaritz A, Spong MW (1996) An experimental comparison of robust control algorithms on a direct drive manipulator. IEEE Trans Control Syst Technol 4(6):627–640
11. Kang BS, Kim SH, Kwak YK, Smith CC (1999) Robust tracking control of a direct drive robot. J Dyn Syst Meas Control 121(2):261–269
12. Lewis FL, Abdallah CT, Dawson DM (2004) Robot manipulator control: theory and practice. Marcel Dekker, Inc., New York
13. Reyes F, Kelly R (2001) Experimental evaluation of model-based controllers on a direct-drive robot arm. Mechatronics 11(3):267–282
14. Safonov MG, Limebeer DJN, Chiang RY (1989) Simplifying the $\mathcal{H}_\infty$ theory via loop shifting, matrix pencil and descriptor concepts. Int J Control 50(6):2467–2488
15. Whitcomb LL, Rizzi AA, Kodistschek DE (1993) Comparative experiments with a new adaptive controller for robot arms. IEEE Trans Rob Autom 9(1):59–70
16. Zhou K, Doyle JC, Glover K (1995) Robust and optimal control. Prentice Hall, Upper Saddle River
17. Zhou K, Doyle JC (1998) Essentials of robust control. Prentice Hall, New York
18. Zhou K, Glover K, Bodenheimer B, Doyle JC (1994) Mixed $\mathcal{H}_2$ and $\mathcal{H}_\infty$ performance objectives I: robust performance analysis. IEEE Trans Autom Control 39(8):1564–1574
19. Zhou K, Glover K, Bodenheimer B, Doyle JC (1994) Mixed $\mathcal{H}_2$ and $\mathcal{H}_\infty$ performance objectives II: optimal control. IEEE Trans Autom Control 39(8):1575–1586

Chapter 3
Nonlinear $\mathcal{H}_\infty$ Control

3.1 Introduction

In this chapter we present a set of nonlinear $\mathcal{H}_\infty$ control methodologies for robot manipulators. In few words, nonlinear $\mathcal{H}_\infty$ control consists in guaranteeing a predefined level of attenuation of the effects of the disturbance in the system output. We deal with two fundamental approaches in this class of controllers: the first is based on game theory and the second is based on linear parameter varying (LPV) techniques.

The $\mathcal{H}_\infty$ control of manipulators based on game theory we are considering in this chapter was developed in [4], taking into account the optimal solution presented in [7]. This approach provides an explicit global solution for the control problem, formulated as a minimax game. The solution proposed in [4] was applied to a fully-actuated experimental manipulator with high inertia in [8]. The approach based on LPV techniques provides a systematic way to design controllers that schedule the varying parameters of the system and achieve the $\mathcal{H}_\infty$ performance [9]. The nonlinear dynamics can be represented as an LPV system in which the parameters are a function of the state, named quasi-LPV representation.

The presentation of both approaches in this chapter is motivated by the fact that the first one provides a constant gain, similar to the results obtained with feedback linearization procedures, while the second one provides a time-varying gain which is a result of the solution of several coupled Riccati inequalities. We investigate the robustness of both approaches through experimental results.

This chapter is organized as follows: Sect. 3.2 presents the quasi-LPV representation of robot manipulators. Section 3.3 presents the $\mathcal{H}_\infty$ control via game theory proposed in [4], as well as a variant based on mixed $\mathcal{H}_2/\mathcal{H}_\infty$ control. Section 3.4 presents the $\mathcal{H}_\infty$ control for LPV systems. Section 3.5 presents guidelines to implement the proposed controllers and the results obtained on the UARM.

A. A. G. Siqueira et al., *Robust Control of Robots*,
DOI: 10.1007/978-0-85729-898-0_3, © Springer-Verlag London Limited 2011

Specific notations used in this chapter include: a matrix C is said to be skew-symmetric when $C = -C^T$; and $\mathcal{L}_2$ will be used to denote the set of bounded-energy signals, i.e., $\mathcal{L}_2(0,\mathrm{T}) = \{w : \int_0^{\mathrm{T}} \|w(\mathrm{t})\|^2 \mathrm{dt} < \infty\}$.

3.2 Quasi-LPV Representation of Robot Manipulators

The dynamic equation of a robot manipulator is given by Eq. 1.1:

$$\tau = M(q)\ddot{q} + C(q,\dot{q})\dot{q} + F(\dot{q}) + G(q). \tag{3.1}$$

In this section we are interested in formulating the manipulator model in terms of nominal parameters, parametric uncertainties and exogenous disturbances. We can divide the parameter matrices $M(q)$, $C(q,\dot{q})$, $F(\dot{q})$, and $G(q)$ into a nominal and a perturbed part:

$$\begin{aligned} M(q) &= M_0(q) + \Delta M(q), \\ C(q,\dot{q}) &= C_0(q,\dot{q}) + \Delta C(q,\dot{q}), \\ F(\dot{q}) &= F_0(\dot{q}) + \Delta F(\dot{q}), \\ G(q) &= G_0(q) + \Delta G(q), \end{aligned}$$

where $M_0(q)$, $C_0(q,\dot{q})$, $F_0(\dot{q})$, and $G_0(q)$ are the nominal matrices, and $\Delta M(q)$, $\Delta C(q,\dot{q})$, $\Delta F(\dot{q})$, and $\Delta G(q)$ are the parametric uncertainties. With these uncertainties and adding a finite energy exogenous disturbance, τ_d, after some algebra, Eq. 3.1 can be rewritten as:

$$\tau + \delta(q,\dot{q},\ddot{q}) = M_0(q)\ddot{q} + C_0(q,\dot{q})\dot{q} + F_0(\dot{q}) + G_0(q), \tag{3.2}$$

with

$$\delta(q,\dot{q},\ddot{q}) = -(\Delta M(q)\ddot{q} + \Delta C(q,\dot{q})\dot{q} + \Delta F(\dot{q}) + \Delta G(q) - \tau_d).$$

In order to express this equation in a form appropriate for the nonlinear control methods we are dealing with in this chapter, the following state tracking error is defined:

$$\tilde{x} = \begin{bmatrix} q - q^d \\ \dot{q} - \dot{q}^d \end{bmatrix} = \begin{bmatrix} \tilde{q} \\ \dot{\tilde{q}} \end{bmatrix}, \tag{3.3}$$

where q^d and $\dot{q}^d \in \Re^n$ are the desired reference trajectory and the corresponding velocity, respectively. The variables q^d, $\dot{q}^d$ and $\ddot{q}^d$, the desired acceleration, are assumed to be within the physical and kinematics limits of the manipulator. The dynamic equation for the state tracking error is given from Eqs. 3.2 and 3.3 as:

$$\dot{\tilde{x}} = A(q,\dot{q})\tilde{x} + Bu + Bw \tag{3.4}$$

with

$$\begin{aligned}
A(q,\dot{q}) &= \begin{bmatrix} 0 & I_n \\ 0 & -M_0^{-1}(q)C_0(q,\dot{q}) \end{bmatrix}, \\
B &= \begin{bmatrix} 0 \\ I_n \end{bmatrix}, \\
u &= M_0^{-1}(q)(\tau - M_0(q)\ddot{q}^d - C_0(q,\dot{q})\dot{q}^d - F_0(\dot{q}) - G_0(q)), \\
w &= M_0^{-1}(q)\delta(q,\dot{q},\ddot{q}).
\end{aligned}$$

The applied torque is given by:

$$\tau = M_0(q)(\ddot{q}^d + u) + C_0(q,\dot{q})\dot{q}^d + F_0(\dot{q}) + G_0(q).$$

Although the matrices $M_0(q)$ and $C_0(q,\dot{q})$ explicitly depend on the joint positions, we can consider them as functions of the position and velocity errors [7]:

$$\begin{aligned}
M_0(q) &= M_0(\tilde{q} + q^d) = M_0(\tilde{x}, t), \\
C_0(q,\dot{q}) &= C_0(\tilde{q} + q^d, \dot{\tilde{q}} + \dot{q}^d) = C_0(\tilde{x}, t).
\end{aligned}$$

Hence, Eq. 3.4 is a quasi-LPV representation for the robot manipulator with $A(\tilde{x}, t)$ representing the state transition matrix.

3.3 $\mathcal{H}_\infty$ Control via Game Theory

In this section, we utilize a classical solution based on game theory for the $\mathcal{H}_\infty$ control problem of a robot manipulator derived from Eq. 3.4. We also discuss a variant based on mixed $\mathcal{H}_2/\mathcal{H}_\infty$ control.

3.3.1 $\mathcal{H}_\infty$ Control

Following [4] and [7], the solution for the $\mathcal{H}_\infty$ control takes into account a state transformation given by:

$$\tilde{z} = \begin{bmatrix} \tilde{z}_1 \\ \tilde{z}_2 \end{bmatrix} = T_0\tilde{x} = \begin{bmatrix} T_1 \\ T_2 \end{bmatrix}\tilde{x} = \begin{bmatrix} I & 0 \\ T_{11} & T_{12} \end{bmatrix}\begin{bmatrix} \tilde{q} \\ \dot{\tilde{q}} \end{bmatrix}, \tag{3.5}$$

where T_{11}, $T_{12} \in \Re^{n\times n}$ are constant matrices to be determined. With this transformation, the dynamic equation of the state tracking error can be rewritten as:

$$\dot{\tilde{x}} = A_T(\tilde{x}, t)\tilde{x} + B_T(\tilde{x}, t)u + B_T(\tilde{x}, t)w, \tag{3.6}$$

with

$$A_T(\tilde{x},t) = T_0^{-1}\begin{bmatrix} -T_{12}^{-1}T_{11} & T_{12}^{-1} \\ 0 & -M_0^{-1}(q)C_0(q,\dot{q}) \end{bmatrix} T_0,$$

$$B_T(\tilde{x},t) = T_0^{-1}\begin{bmatrix} 0 \\ M_0^{-1}(q) \end{bmatrix},$$

$$u = M_0(q)T_2\dot{\tilde{x}} + C_0(q,\dot{q})T_2\tilde{x},$$

$$w = M_0(q)T_{12}M_0^{-1}(q)\delta(q,\dot{q},\ddot{q}).$$

The relationship between the applied torques and the control input is given by:

$$\tau = M_0(q)\ddot{q} + C_0(q,\dot{q})\dot{q} + F_0(\dot{q}) + G_0(q), \tag{3.7}$$

with

$$\ddot{q} = \ddot{q}^d - T_{12}^{-1}T_{11}\dot{\tilde{x}} - T_{12}^{-1}M_0^{-1}(q)\big(C_0(q,\dot{q})B^T T_0\tilde{x} - u\big). \tag{3.8}$$

The objective of the robust control considered in this section is to attenuate the effects of the disturbance w on the position and velocity of the manipulator joints through the state feedback control $u = F(\tilde{x})\tilde{x}$. With this strategy in mind, and subject to the tracking error dynamics, the following performance criterion, which includes a desired disturbance attenuation level γ, is proposed in [4]:

$$\min_{u(\cdot)\in\mathcal{L}_2}\ \max_{0\neq w(\cdot)\in\mathcal{L}_2} \frac{\int_0^\infty \big(\frac{1}{2}\tilde{x}^T(t)Q\tilde{x}(t) + \frac{1}{2}u^T(t)Ru(t)\big)dt}{\int_0^\infty \big(\frac{1}{2}w^T(t)w(t)\big)dt} \leq \gamma^2, \tag{3.9}$$

where Q and R are positive definite symmetric weighting matrices and $\tilde{x}(0) = 0$. This performance criterion represents the classical minimax optimization problem with weighting matrices introduced in the output and in the control input terms. According to game theory, a solution to this problem can be found if there exists a continuously differentiable Lyapunov function $V(\tilde{x},t)$ that satisfies the following Bellman-Isaacs equation [2]:

$$-\frac{\partial V(\tilde{x},t)}{\partial t} = \min_{u(\cdot)}\max_{w(\cdot)}\left\{ L(\tilde{x},u,w) + \left(\frac{\partial V(\tilde{x},t)}{\partial \tilde{x}}\right)^T \dot{\tilde{x}} \right\},$$

with terminal condition $V(\tilde{x}(\infty),\infty) = 0$ and $L(\tilde{x},u,w) = \frac{1}{2}\tilde{x}^T(t)Q\tilde{x}(t) + \frac{1}{2}u^T(t)$ $Ru(t) - \frac{1}{2}\gamma^2 w^T(t)w(t)$. Selecting a Lyapunov function of the form:

$$V(\tilde{x},t) = \frac{1}{2}\tilde{x}^T P(\tilde{x},t)\tilde{x}, \tag{3.10}$$

where $P(\tilde{x},t)$ is a positive definite symmetric matrix for all $\tilde{x}$ and t, the Bellman-Isaacs equation becomes the following Riccati equation:

$$\dot{P}(\tilde{x},t) + P(\tilde{x},t)A_T(\tilde{x},t) + A_T(\tilde{x},t)P(\tilde{x},t) + P(\tilde{x},t)B_T(\tilde{x},t)\left(R^{-1} - \frac{1}{\gamma^2}I\right)B_T(\tilde{x},t)P(\tilde{x},t) + Q = 0. \quad (3.11)$$

With an appropriate choice of the matrix $P(\tilde{x},t)$ and by use of the skew-symmetric matrix $C_0(q,\dot{q}) - \frac{1}{2}\dot{M}_0(q,\dot{q})$, the Riccati equation can be simplified to an algebraic matrix equation. In [4, 7], the $P(\tilde{x},t)$ selected as solution for this problem is defined as:

$$P(\tilde{x},t) = T_0^T \begin{bmatrix} K & 0 \\ 0 & M_0(\tilde{x},t) \end{bmatrix} T_0, \quad (3.12)$$

where K is a positive definite symmetric constant matrix. The simplified algebraic equation is given as:

$$\begin{bmatrix} 0 & K \\ K & 0 \end{bmatrix} - T_0^T B\left(R^{-1} - \frac{1}{\gamma^2}I\right)B^T T_0 + Q = 0. \quad (3.13)$$

The $\mathcal{H}_\infty$ robust control is obtained through the following simple and elegant procedure:

$$u^* = -R^{-1}B^T T_0 \tilde{x}. \quad (3.14)$$

One can note that the resulting control input u is actually a static gain. The terminal condition is satisfied for the choice of $P(\tilde{x},t)$ as in Eq. 3.12 [4]. Then, to solve the algebraic Eq. 3.13, we can adopt as solution the following T_0 and K matrices:

$$T_0 = \begin{bmatrix} I & 0 \\ R_1^T Q_1 & R_1^T Q_2 \end{bmatrix} \quad (3.15)$$

and

$$K = \frac{1}{2}\left(Q_1^T Q_2 + Q_2^T Q_1\right) - \frac{1}{2}\left(Q_{21}^T + Q_{12}\right),$$

with the conditions $K > 0$ and $R < \gamma^2 I$. Matrix R_1 is the result of the Cholesky factorization:

$$R_1^T R_1 = \left(R^{-1} - \frac{1}{\gamma^2}I\right)^{-1} \quad (3.16)$$

and the positive definite symmetric matrix Q is factored as

$$Q = \begin{bmatrix} Q_1^T Q_1 & Q_{12} \\ Q_{12}^T & Q_2^T Q_2 \end{bmatrix}. \tag{3.17}$$

Some remarks can be made about the selection of the weighting matrices Q_1, Q_2, and R and the attenuation level γ:

- There exists a compromise between the parameter γ and the weighting matrix R: firstly R is selected and then γ is adjusted according to the constraint $R < \gamma^2 I$.
- As $\gamma \to \infty$ the resulting controller approximates the $\mathcal{H}_2$ controller, which does not guarantee disturbance attenuation.
- In order to satisfy the conditions $K > 0$ and $R < \gamma^2 I$, and to obtain a feasible solution to the Cholesky factorization, it is easier to select the weighting matrices Q_1, Q_2, and R as diagonal matrices, and $Q_{12} = 0$. Let $Q_1 = \eta_1 I_n$, $Q_2 = \eta_2 I_n$, and $R = r I_n$. With these choices and considering Eqs. 3.7, 3.8, and 3.14, the applied torque becomes

$$\begin{aligned} \tau = {} & M_0(q)\left(\ddot{q}^d - \frac{\eta_1}{\eta_2}\dot{\tilde{x}}_2\right) + C_0(q,\dot{q})\left(\dot{q}^d - \frac{\eta_1}{\eta_2}\tilde{x}_2\right) \\ & + F_0(\dot{q}) + G_0(q) - \frac{1}{r}\begin{bmatrix} I_n & \frac{\eta_1}{\eta_2} I_n \end{bmatrix}\tilde{x}. \end{aligned} \tag{3.18}$$

- If off-diagonal elements are added to the matrix Q_2, even when Q_2 remains positive definite, matrix T_{12}^{-1} may contain negative elements. In this case, positive feedback may appear in the control law (Eq. 3.7).

3.3.2 Mixed $\mathcal{H}_2/\mathcal{H}_\infty$ Control

The mixed $\mathcal{H}_2/\mathcal{H}_\infty$ nonlinear control aims to minimize a quadratic cost while attenuating disturbances. This problem was solved in [3] based on game theory. In the mixed case, the $\mathcal{H}_2$ optimal control must be obtained while ensuring that the $\mathcal{H}_\infty$ performance criterion is satisfied. Given a desired disturbance level $\gamma > 0$ and weighting matrices Q_1, Q_2, and R, the mixed $\mathcal{H}_2/\mathcal{H}_\infty$ control problem is solved if there exist a controller u such that the $\mathcal{H}_2$ optimal cost

$$\min_{u(t)} J_2(u,w) \tag{3.19}$$

can be achieved under the $\mathcal{H}_\infty$ constraint:

$$\max_{w(t)\in\mathcal{L}_2[0,t_f]} J_1(u,w) \le \tilde{x}^T(0) P \tilde{x}(0), \tag{3.20}$$

with

$$J_2(u,w) = \tilde{x}^T(t_f)Q_{2f}\tilde{x}(t_f) + \int_0^{t_f} (\tilde{x}^T(t)Q_2\tilde{x}(t) + u^T(t)Ru(t))dt$$

and

$$J_1(u,w) = \tilde{x}^T(t_f)Q_{1f}\tilde{x}(t_f) + \int_0^{t_f} (\tilde{x}^T(t)Q_1\tilde{x}(t) + u^T(t)Ru(t))dt$$
$$-\gamma^2 \int_0^{t_f} w^T(t)w(t)dt,$$

where $P = P^T > 0$, $Q_{1f} = Q_{1f}^T > 0$, and $Q_{2f} = Q_{2f}^T > 0$. The solution for this problem is given by the following coupled algebraic equations:

$$\begin{bmatrix} 0 & K_1 \\ K_1 & 0 \end{bmatrix} - T_0^T B\left(R^{-1} - \frac{1}{\gamma^2}I_n\right)B^T T_0 + Q_1 = 0 \quad (3.21)$$

and

$$\begin{bmatrix} 0 & K_2 \\ K_2 & 0 \end{bmatrix} - T_0^T B\left(R^{-1} - \frac{2}{\gamma^2}I_n\right)B^T T_0 + Q_2 = 0, \quad (3.22)$$

where $B = [I_n \ 0]^T$. The optimal control and the worst-case disturbance are given by:

$$u^* = -R^{-1}B^T T_0\tilde{x}; \quad w^* = \frac{1}{\gamma^2}B^T T_0\tilde{x}.$$

To solve Eqs. 3.21 and 3.22 some constraints are required to compute the matrices Q_1, Q_2, and R. Subtracting Eq. 3.22 from 3.21 yields:

$$\begin{bmatrix} 0 & K_1 - K_2 \\ K_1 - K_2 & 0 \end{bmatrix} - \frac{1}{\gamma^2}T_0^T BB^T T_0 + Q_1 - Q_2 = 0.$$

Since $(1/\gamma^2)T^T BB^T T$ is positive definite, the constraint, $Q_1 > Q_2 > 0$, must hold. The positive definite symmetric matrices Q_1 and ΔQ can be factored as:

$$Q_1 = \begin{bmatrix} Q_{11}^T Q_{11} & Q_{12} \\ Q_{12}^T & Q_{22}^T Q_{22} \end{bmatrix}, \quad \Delta Q = \begin{bmatrix} \Delta Q_{11}^T \Delta Q_{11} & \Delta Q_{12} \\ \Delta Q_{12}^T & \Delta Q_{22}^T \Delta Q_{22} \end{bmatrix}$$

where $\Delta Q = Q_1 - Q_2$. The solutions of Eqs. 3.21 and 3.22 are given by

$$T_0 = \begin{bmatrix} I_n & 0 \\ \gamma\Delta Q_{11} & \gamma\Delta Q_{22} \end{bmatrix}, \quad (3.23)$$

$$K_1 = \frac{1}{2}\left(Q_{11}^T Q_{22} + Q_{22}^T Q_{11}\right) - \frac{1}{2}\left(Q_{12}^T + Q_{12}\right), \tag{3.24}$$

and

$$K_2 = \frac{1}{4}\left(Q_{11}^T Q_{22} + Q_{22}^T Q_{11}\right) - \frac{1}{4}\left(Q_{12}^T + Q_{12}\right), \tag{3.25}$$

where K_1 and K_2 are positive definite matrices $(Q_{11}^T Q_{22} + Q_{22}^T Q_{11} > Q_{12}^T + Q_{12})$. To guarantee that these two coupled equations are solvable, the matrix R must be of the form:

$$R = \gamma^2[I_n + (Q_{11}\Delta Q_{11}^{-1})^T(Q_{11}\Delta Q_{11}^{-1})]^{-1}. \tag{3.26}$$

3.4 $\mathcal{H}_\infty$ Control via Linear Matrix Inequalities

In this section we present another class of nonlinear $\mathcal{H}_\infty$ controllers. The designs we consider here are based on state and on output feedback approaches. We deal with linear parameter-varying (LPV) systems whose control procedure is synthesized in terms of linear matrix inequalities (LMIs). It is important to emphasize an interesting difference between this approach and the solution we presented in Sect. 3.3, based on game theory. The solutions obtained through game theory are based on algebraic Riccati equations in which the controller gain is fixed, independently of parameter variations in the manipulator. In the solutions provided in this section we solve a convex optimization problem through a set of LMIs and as a by-product we obtain a time-varying controller gain. The advantage of this procedure is that we can incorporate the parameters' derivatives in the derivation of the controller. The optimization problem in this approach is based on the following nonlinear system with exogenous disturbances $w \in \Re^p$, control inputs $u \in \Re^m$, and output variables $z \in \Re^q$:

$$\begin{aligned} \dot{x} &= f(x) + b_1(x)w + b_2(x)u, \\ z &= h(x) + d_1(x)w + d_2(x)u, \end{aligned} \tag{3.27}$$

where $f(0) = 0$ and $h(0) = 0$, and $x \in \Re^n$ are state variables. We assume that $f(\cdot)$, $b_i(\cdot)$, $h(\cdot)$, $d_i(\cdot)$ are continuously differentiable functions. The performance of the system (3.27) is defined by adjusting the controller to guarantee that the $\mathcal{L}_2$ gain between the disturbance and the output is satisfied, that is,

$$\int_0^T \|z(t)\|^2 dt \le \gamma^2 \int_0^T \|w(t)\|^2 dt, \tag{3.28}$$

for all $T \geq 0$ and all $w \in \mathcal{L}_2(0,\mathrm{T})$ with the system starting from $x(0) = 0$. Note that condition (3.28) is equivalent to the quadratic functional proposed in Eq. 3.9. The parameter γ assumes an equivalent role to that defined in the standard $\mathcal{H}_\infty$ criteria for linear systems. In virtue of this we can consider that this functional defines an equivalent nonlinear $\mathcal{H}_\infty$ control problem.

3.4.1 State Feedback $\mathcal{H}_\infty$ Control for LPV Systems

For the control system we consider in this section, we assume that positions and velocities of the manipulator joints are measured properly through sensors. Consider the following LPV model:

$$\begin{bmatrix} \dot{x} \\ y \\ z \end{bmatrix} = \begin{bmatrix} A(\rho(x)) & B_1(\rho(x)) & B_2(\rho(x)) \\ C_1(\rho(x)) & 0 & 0 \\ C_2(\rho(x)) & 0 & I \end{bmatrix} \begin{bmatrix} x \\ w \\ u \end{bmatrix}, \qquad (3.29)$$

where x is the state vector, u is the control input vector, w is the external input vector, y and z are the output variables, and $\rho(x)$ is the parameter varying vector. Assume that the underlying parameter $\rho(x)$ varies in the allowable set:

$$F_P^\nu = \{\rho(x) \in C^1(\Re^n, \Re^m) : \rho(x) \in P, |\dot{\rho}_i| \leq \nu_i\}, \qquad (3.30)$$

for $i = 1,\ldots,k$, where $P \subset \Re^m$ is a compact set, $\nu = [\nu_1 \cdots \nu_m]^T$ with $\nu_i \geq 0$ and $\underline{\nu}_i(\rho) \leq \dot{\rho}_i \leq \overline{\nu}_i(\rho)$, $i = 1,\ldots,m$. We denote by $C^1(\Re^n, \Re^m)$ the set of continuously differentiable functions that map $\Re^n$ to $\Re^m$. The state feedback control problem considered here aims at finding a continuous function $F(\rho)$ such that the closed loop system has an $\mathcal{L}_2$ gain less than or equal to γ under a state feedback law $u = F(\rho)x$. According to [9],

if there exists a continuously differentiable matrix function $X(\rho(x)) > 0$ that satisfies:

$$\begin{bmatrix} E(\rho(x)) & X(\rho(x))C_1^T(\rho(x)) & B_1(\rho(x)) \\ C_1(\rho(x))X(\rho(x)) & -I & 0 \\ B_1^T(\rho(x)) & 0 & -\gamma^2 I \end{bmatrix} < 0, \qquad (3.31)$$

where

$$E(\rho(x)) = -\sum_{i=1}^{m} \underline{\overline{\nu}}_i(\rho) \frac{\partial X(\rho(x))}{\partial \rho_i} - B_2(\rho(x))B_2^T(\rho(x))$$
$$+ \widehat{A}(\rho(x))X(\rho(x)) + X(\rho(x))\widehat{A}(\rho(x))^T$$

and $\widehat{A}(\rho(x)) = A(\rho(x)) - B_2(\rho(x))C_2(\rho(x))$, then the closed loop system has $\mathcal{L}_2$ gain $\leq \gamma$ under the state feedback control law

$$u(t) = -(B_2(\rho(x))X^{-1}(\rho(x)) + C_2(\rho(x)))x(t).$$

Under the assumption that the underlying parameter $\rho(x)$ varies within the allowable set F_P^v, we can combine the effect of the positions and velocities derivatives of the manipulator through the summation $\sum_{i=1}^m \underline{\overline{v}}_i(\rho)$—representing that every combination of $\overline{v}_i(\rho)$ and $\underline{v}_i(\rho)$ should be included in the inequality. Hence, Eq. 3.31 actually represents 2^m inequalities. The solution of this set of LMIs characterizes an infinitesimal convex optimization problem, which is hard to solve numerically. With some approximations, a practical scheme was developed in [9] to compute these LMIs based on basis functions related to the $X(\rho(x))$ and on the grid of the parameter set P. First we need to choose a set of C^1 basis functions $\{\phi_i(\rho)\}_{i=1}^M$ for $X(\rho(x))$ in order to rewrite it as:

$$X(\rho(x)) = \sum_{i=1}^{M} \phi_i(\rho(x))X_i, \tag{3.32}$$

where $X_i \in S^{n\times n}$ is the coefficient matrix for $\phi_i(\rho(x))$. Applying this matrix $X(\rho(x))$ in Eq. 3.31, the constraint turns into a LMI in terms of the matrix variables $\{X_i\}_{i=1}^M$, when the parameter $\rho(x)$ is fixed, and we can define the following

Optimization problem:

$$\min_{\{X_i\}_{i=1}^M} \gamma^2$$

subject to

$$\begin{bmatrix} E^*(\rho) & \sum_{j=1}^M \phi_j(\rho)X_jC_1^T(\rho) & B_1(\rho) \\ C_1(\rho)\sum_{j=1}^M \phi_j(\rho)X_j & -I & 0 \\ B_1^T(\rho) & 0 & -\gamma^2 I \end{bmatrix} < 0,$$

$$\sum_{j=1}^{M} \phi_j(\rho)X_j > 0, \tag{3.33}$$

where

$$E^*(\rho) = -\sum_{i=1}^{m} \underline{\overline{v}}_i(\rho)\frac{\partial X(\rho(x))}{\partial \rho_i} - B_2(\rho)B_2^T(\rho) + \sum_{j=1}^{M} \phi_j(\rho)(\widehat{A}(\rho)X_j + X_j\widehat{A}(\rho)^T).$$

To solve this infinite dimensional optimization problem, we can grid the parameter set P along L points $\{\rho_k\}_{k=1}^{L}$ in each dimension. Since Eq. 3.31 consists of 2^m entries, a total of $(2^m+1)L^m$ matrix inequalities in terms of the matrices $\{X_i\}$ should be solved. This computation has a few obvious limitations. The number of parameters and the number of points L should be chosen such that the solution is reached in a feasible number of iterations. A lower limit for L so that it is possible to find a global solution for all LMIs was proposed in [9]. Another problem is the lack of guidance in choosing the basis functions ϕ_i. In terms of the robotic applications we are considering in this chapter, an useful procedure is to define these functions according to the variables that define the dynamics of the robot manipulator—for example, $\sin(q_i)$ and $\cos(q_i)$ which appear in the inertial, centripetal, and Coriolis matrices. Section 3.5 in this chapter presents some examples.

3.4.2 Output Feedback $\mathcal{H}_\infty$ LPV Control

In this section, we present an approach for controlling robotic manipulators based on output feedback techniques and on LPV systems. In this formulation, the controller aims to stabilize the closed loop system while guaranteeing that the $\mathcal{L}_2$ gain between the disturbance and the output of the system is bounded by an attenuation level γ. Output feedback control of robotic applications is interesting from the point of view of economy of sensors. The designer can avoid, for example, installing joint velocity sensors and rely instead on position sensors only. The LPV-based representation in this control strategy is justified with the same arguments used in the state-feedback control aforementioned: it is close to the Lagrange–Euler equations, in which positions and velocities of the robot manipulator are part of the parameter matrices of the model. To apply the control techniques presented in [1], the robot manipulator needs to be represented in the following form:

$$\begin{aligned}
\dot{x} &= A(\rho(x))x + B_1(\rho(x))w + B_2(\rho(x))u, \\
z &= C_1(\rho(x))x + D_{11}(\rho(x))w + D_{12}(\rho(x))u, \\
y &= C_2(\rho(x))x + D_{21}(\rho(x))w,
\end{aligned} \tag{3.34}$$

where $\rho(x) = [\rho_1(x), \ldots, \rho_N(x)]^T$ belongs to a convex space P, and $\rho_i(x)$, $i = 1, \ldots, N$, are the time-varying parameters satisfying $|\dot{\rho}_i(x)| \leq \nu_i$ with $\nu_i \geq 0$, $i = 1, \ldots, N$, the bounds of the parameter variation rates. Consider as system disturbances the desired position, q^d, and the combined torque disturbance, δ, that is: $w = [\delta^T \ (q^d)^T]^T$. The system outputs, z, are the position error, $q^d - q$, and the control input, u. The control output is the position error, $y = q^d - q$, since only the position is measured directly. In this case, the manipulator can be described by (3.34) with:

$$A(\rho(x)) = A(q,\dot{q}), \qquad B_1(\rho(x)) = [B(q)\ 0], \qquad B_2(\rho(x)) = B(q),$$
$$C_1(\rho(x)) = \begin{bmatrix} 0 & -I \\ 0 & 0 \end{bmatrix}, \qquad C_2(\rho(x)) = [0\ -I], \qquad D_{11}(\rho(x)) = \begin{bmatrix} 0 & I \\ 0 & 0 \end{bmatrix},$$
$$D_{12}(\rho(x)) = [0\ I]^T, \qquad D_{21}(\rho(x)) = [0\ I], \qquad D_{22}(\rho(x)) = 0,$$

where $A(q,\dot{q})$ and $B(q)$ are obtained from (3.4). In [1], the authors propose two $\mathcal{H}_\infty$ control techniques to solve this control problem. We use the approach named *projected characterization*, which is based on fundamental results developed in [5], to control a manipulator based on the LPV model (3.34). The controller dynamics are defined as:

$$\begin{bmatrix} \dot{x}_K \\ u \end{bmatrix} = \begin{bmatrix} A_K(\rho(x),\dot{\rho}(x)) & B_K(\rho(x),\dot{\rho}(x)) \\ C_K(\rho(x),\dot{\rho}(x)) & D_K(\rho(x),\dot{\rho}(x)) \end{bmatrix} \begin{bmatrix} x_K \\ y \end{bmatrix}. \tag{3.35}$$

To obtain the controller we must first solve a set of LMIs in $X(\rho(x))$ and $Y(\rho(x))$ minimizing γ according to the following

Projected characterization algorithm:

$$\left[\begin{array}{c|c} N_X & 0 \\ \hline 0 & I \end{array}\right]^T \left[\begin{array}{cc|c} \dot{X} + XA + A^{TX} & XB_1 & C_1^T \\ B_1^T X & -\gamma I & D_{11}^T \\ \hline B_1 & D_{11} & -\gamma I \end{array}\right] \left[\begin{array}{c|c} N_X & 0 \\ \hline 0 & I \end{array}\right] < 0, \tag{3.36}$$

$$\left[\begin{array}{c|c} N_Y & 0 \\ \hline 0 & I \end{array}\right]^T \left[\begin{array}{cc|c} -\dot{Y} + YA^T + AY & YC_1^T & B_1 \\ C_1 Y & -\gamma I & D_{11} \\ \hline B_1^T & D_{11}^T & -\gamma I \end{array}\right] \left[\begin{array}{c|c} N_Y & 0 \\ \hline 0 & I \end{array}\right] < 0, \tag{3.37}$$

$$\begin{bmatrix} X & I \\ I & Y \end{bmatrix} > 0, \tag{3.38}$$

where N_X and N_Y designate any bases of the null spaces of $[C_2\ D_{21}]$ and $[B_2^T\ D_{12}^T]$, respectively. After finding X and Y, compute D_K to satisfy:

$$\sigma_{\max}(D_{11} + D_{12} D_K D_{21}) < \gamma \tag{3.39}$$

and set $D_{cl} := D_{11} + D_{12} D_K D_{21}$; compute $\widehat{A}_K$, $\widehat{B}_K$ and $\widehat{C}_K$ through

$$\begin{bmatrix} 0 & D_{21} & 0 \\ D_{21}^T & -\gamma I & D_{cl}^T \\ 0 & D_{cl} & -\gamma I \end{bmatrix} \begin{bmatrix} \widehat{B}_K^T \\ \star \end{bmatrix} = -\begin{bmatrix} C_2 \\ B_1^T X \\ C_1 + D_{12} D_K C_2 \end{bmatrix}, \tag{3.40}$$

$$\begin{bmatrix} 0 & D_{12}^T & 0 \\ D_{12} & -\gamma I & D_{cl} \\ 0 & D_{cl}^T & -\gamma I \end{bmatrix} \begin{bmatrix} \widehat{C}_K \\ \star \end{bmatrix} = -\begin{bmatrix} B_2^T \\ C_1 Y \\ (B_1 + B_2 D_K D_{21})^T \end{bmatrix}, \tag{3.41}$$

$$\widehat{A}_K = -(A + B_2 D_K C_2)^T + [XB_1 + \widehat{B}_K D_{21}(C_1 + D_{12} D_K C_2)^T]$$
$$\times \begin{bmatrix} -\gamma I & D_{cl}^T \\ D_{cl} & -\gamma I \end{bmatrix}^{-1} \begin{bmatrix} (B_1 + B_2 D_K D_{21})^T \\ C_1 Y + D_{12}\widehat{C}_K \end{bmatrix}, \tag{3.42}$$

where ☆ means that the respective entries are not important to the computation of the controller. N and M are solved through the factorization problem $I - XY = NM^T$. Finally, compute A_K, B_K, and C_K as:

$$A_K = N^{-1}(X\dot{Y} + \widehat{A}_K - X(A - B_2 D_K C_2)Y - \widehat{B}_K C_2 Y - XB_2 \widehat{C}_K)M^{-T},$$
$$B_K = N^{-1}(\widehat{B}_K - XB_2 D_K),$$
$$C_K = (\widehat{C}_K - D_K C_2 Y)M^{-T}.$$

Note that the matrices in this algorithm depend on $\rho(x)$; this dependency was omitted for convenience. The LMI problem defined through Eqs. 3.36–3.38 is infinite-dimensional, since the parameter vector $\rho(x)$ varies continuously. To solve this problem, we can divide the parameter space, P, in several points. The variables $X(\rho(x))$ and $Y(\rho(x))$ will be a solution if both satisfy the LMIs for all points. There is no systematic rule that defines how $X(\rho(x))$ and $Y(\rho(x))$ vary upon $\rho(x)$, although this problem is usually solved using basis functions to describe the unknown matrices, which lead them to be written as:

$$X(\rho(x)) = \sum_{i=1}^{M} \phi_i(\rho(x))X_i,$$
$$Y(\rho(x)) = \sum_{i=1}^{M} \psi_i(\rho(x))Y_i, \tag{3.43}$$

where $\{\phi_i(\rho(x))\}_{i=1}^M$ and $\{\psi_i(\rho(x))\}_{i=1}^M$ are differentiable functions of $\rho(x)$ (see [1] for details). Note that to obtain the best performance with this controller, the choice of v_i (in order to guarantee $|\dot{\rho}_i(x)| \le v_i$) should be checked *a posteriori*.

3.5 Examples

In this section we present the application of the $\mathcal{H}_\infty$ and $\mathcal{H}_2/\mathcal{H}_\infty$ control approaches discussed in this chapter to the UARM. We present guidelines to select the weighting matrices when using game theory and to select basis functions when using linear matrices inequalities.

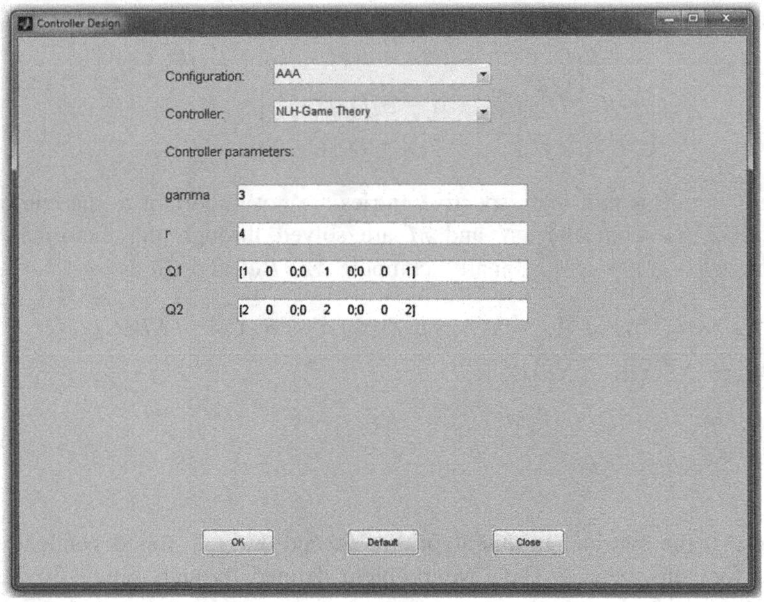

Fig. 3.1 Controller Design box for $\mathcal{H}_\infty$ control via game theory

3.5.1 Design Procedures

The implementation of the $\mathcal{H}_\infty$ controller via game theory is relatively simple. The control design parameters can be selected in the Control Environment for Robot Manipulators through the Controller Design box (Fig. 3.1), by pressing the *Controller Design* button and selecting `NLH-Game Theory` in the menu `Controller`. The following parameters can be selected:

- *gamma*: Defines the value of the attenuation level γ.
- *r*: Defines the positive definite symmetric weighting matrix R. For simplicity, R is considered of the form rI_n.
- *Q1*: Defines the positive definite symmetric weighting matrix Q_1.
- *Q2*: Defines the positive definite symmetric weighting matrix Q_2.

Matrices T_0, T_{11}, and T_{12} are obtained following the steps described in Sect. 3.3. The Cholesky factorization, Eq. 3.16, is computed by the Matlab function $Y = chol(X)$. If X is positive definite, this function results in an upper triangular matrix Y so that $Y^T Y = X$. An error message is printed if X is not positive definite. For the proposed controller, the user must select $r < \gamma^2$ to satisfy this condition. From Eqs. 3.3 and 3.17, note that matrices Q_1 and Q_2 are related with the joint velocity and position errors, respectively. Also, from the solution of the algebraic Eq. 3.15 and the control action (3.14), matrices Q_1 and Q_2 work

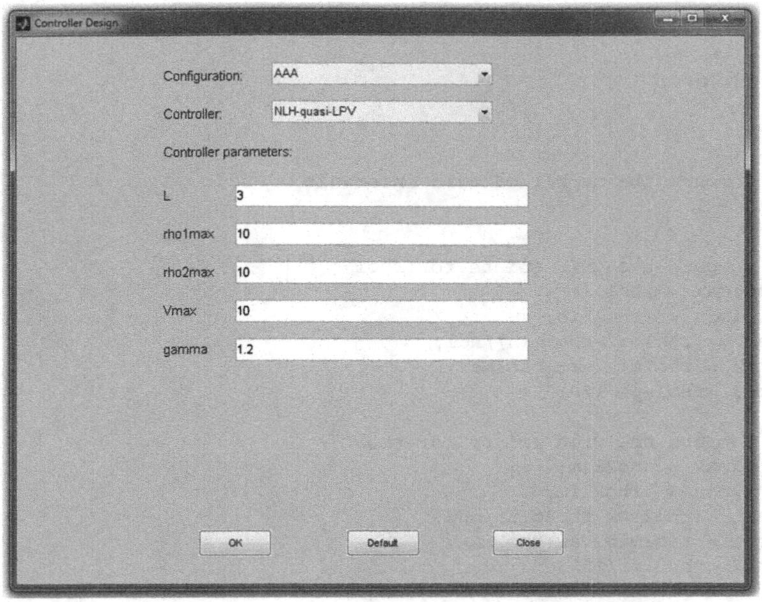

Fig. 3.2 Controller Design box for $\mathcal{H}_\infty$ controller via quasi-LPV representation

directly in the joint velocity and position corrections, acting like a PD controller. On the other hand, matrix R works in both position and velocity errors through the matrix R_1.

For the $\mathcal{H}_\infty$ controller via quasi-LPV representation, the user can select the following control parameters through the Control Design box shown in Fig. 3.2:

- *L*: Defines the number of points in the grid of the parameter set P.
- *rho1max*: Defines the maximum absolute value for the first element of the parameter vector ρ defined in Sect. 3.5.3. The parameter range is defined as $\rho_1 \in \left[-\rho_{1_{max}}, \rho_{1_{max}}\right]$.
- *rho2max*: Defines the maximum absolute value for the second element of parameter ρ.
- $V_{\max}$: Defines the maximum value for the variation rates of the parameters ρ_i. It is considered the same value for all parameters.
- *gamma*: Defines the value for the attenuation level γ.

The $\mathcal{H}_\infty$ controller via quasi-LPV representation is designed via the following steps. First, the dynamic matrices $A(\rho)$, $B_1(\rho)$, $B_2(\rho)$, $C_1(\rho)$, and $C_2(\rho)$ are computed for each point of the grid defined by the parameter set. The number of points in the grid is specified by the control parameters *L*, *rho1lim* and *rho2lim*. The following Matlab code performs the computation of the grid points and the dynamic matrices; this code is found in the directory `CERob\Underactuated Simulator`.

File: ab_lpv.m

```
...

% Compute the number of grid intervals
nd1 = L - 1;
nd2 = L - 1;

% Compute the grid points for rho1
rho1max = rho1lim
rho1min = -rho1lim;
var = (rho1max-rho1min)/nd1;
rho1 = rho1min:var:rho1max;
rho1 =rho1*pi/180;

% Compute the grid points for rho2
rho2max = rho2lim;
rho2min = -rho2lim;
var = (rho2max-rho2min)/nd2;
rho2 = rho2min:var:rho2max;
rho2 =rho2*pi/180;

n1 = size(rho1,2);
n2 = size(rho2,2);

% Define the bound for parameter variation
v = [vmax*pi/180 vmax*pi/180];
omega = [v(1)/2 v(1)/2 v(1)/2];

% Compute dynamic matrices of the quasi-LPV model
B1 = [zeros(3); eye(3)];
B2 = [zeros(3); eye(3)];
C1 = eye(6);
C2 = zeros(3,6);

p = 1;
for i = 1:n1
    for j = 1:n2

        % Joint position
        theta = [0; rho1(i); rho2(j)];

        % Robot dynamics
        [M,C] = pr3l_nlh(theta,omega,m,l,lc,I,atr);
        W = inv(M);

        % Dynamic matrices of the quasi-LPV model
        A(:,:,p) = [zeros(3) eye(3); zeros(3) -W*C];
```

```
            % Compute the basis functions values
            f_1(p)  = 1;
            f_2(p)  = cos(theta(2));
            f_3(p)  = cos(theta(3));

            % Define the derivative of the basis functions
            dfdx_v(:,:,p)  =  [0                  0;
                              -sin(theta(2))   0;
                              0                  -sin(theta(3))];
            p = p + 1;
        end
    end
```

The second step in the controller design procedure is the computation of the matrices X_i, which define the controller itself. These matrices satisfy the set of LMIs defined by Eq. 3.32. The number of variables X_i is specified by the number of basis functions $\phi_i(\rho(x))$ defined in the code above. For the examples shown in this book, three basis functions are used: $\phi_1(\rho(x)) = 1$, which corresponds to the fixed component of $X(\rho)$; $\phi_2(\rho(x)) = \cos(\tilde{q}_2)$ and $\phi_3(\rho(x)) = \cos(\tilde{q}_3)$, which correspond to the variable component of $X(\rho)$ and are in accordance with the practical procedure mentioned in Sect. 3.4.1. The solution of the LMI problem is found using the LMI Toolbox developed by Gahinet et al. [6]. The following Matlab code performs the definition of the LMIs and the computation of the controller solution.

File: lpv_3AAA.m

```
...

% Define the size of the dynamic matrices
nx =size(A,2);

% --------------------
%         LMIs
% --------------------

setlmis([]);

% Define the LMI variables
X1 = lmivar(1,[nx 1]);
X2 = lmivar(1,[nx 1]);
X3 = lmivar(1,[nx 1]);
GAMMA2 = lmivar(1,[nx/2 0]);

% Define the LMIs
for i=1:npts
    %define the (rho)
    A_h = A(:,:,i) - B2*C2;
```

```
% Define the basis functions and their derivatives
f1 = f_1(i);
f2 = f_2(i);
f3 = f_3(i);
dfdx = dfdx_v(:,:,i);

% Compute the LMI terms
for l=1:4

    BRL=newlmi;
    lmiterm([BRL 1 1 X1],-(aux(1,l)*v(1)*dfdx(1,1) ...
     + aux(2,l)*v(2)*dfdx(1,2)),1);
    lmiterm([BRL 1 1 X2],-(aux(1,l)*v(1)*dfdx(2,1) ...
     + aux(2,l)*v(2)*dfdx(2,2)),1);
    lmiterm([BRL 1 1 X3],-(aux(1,l)*v(1)*dfdx(3,1) ...
     + aux(2,l)*v(2)*dfdx(3,2)),1);
        lmiterm([BRL 1 1 X1],f1*A_h,1,'s');
        lmiterm([BRL 1 1 X2],f2*A_h,1,'s');
        lmiterm([BRL 1 1 X3],f3*A_h,1,'s');
        lmiterm([BRL 1 1 0],-B2*B2');
        lmiterm([BRL 1 2 X1],f1,C1');
        lmiterm([BRL 1 2 X2],f2,C1');
        lmiterm([BRL 1 2 X3],f3,C1');
        lmiterm([BRL 1 3 0],B1);
        lmiterm([BRL 2 2 0],-1);
        lmiterm([BRL 2 3 0],0);
        lmiterm([BRL 3 3 GAMMA2],-1,1);

    end

    % Compute the 2nd LMI
    Xpos = newlmi;
    lmiterm([-Xpos 1 1 X1],f1,1);
    lmiterm([-Xpos 1 1 X2],f2,1);
    lmiterm([-Xpos 1 1 X3],f3,1);
end

LMIs = getlmis;

% Define the variable to be minimized (GAMMA2)
ndc = decnbr(LMIs);
c =zeros(ndc,1);
c(ndc) = 1;

% Solve the set of LMIS
[copt,xopt] = mincx(LMIs,c,[1e-2 300 1e9 0 0]);

% Compute the controller matrices
X1f = dec2mat(LMIs,xopt,X1);
X2f = dec2mat(LMIs,xopt,X2);
X3f = dec2mat(LMIs,xopt,X3);

% Compute the optimal solution for GAMMA
gamma = sqrt(copt);
```

3.5.2 Controller Design via Game Theory

For all experimental results of the controllers based on game theory we provide in this subsection we adopt the initial position $q(0) = [0^\circ\ 0^\circ\ 0^\circ]^T$ and desired final position $q(T) = [-20^\circ\ 30^\circ\ -30^\circ]^T$, where the vector $T = [3.4\ 4.0\ 4.0]$ s contains the trajectory duration time for each joint. Exogenous disturbances of the following form are introduced at $t_d = 1.5$ s:

$$\tau_d = \begin{bmatrix} -0.08e^{-2(t-t_d)}\sin(2\pi(t-t_d)) \\ 0.04e^{-2(t-t_d)}\sin(2\pi(t-t_d)) \\ -0.02e^{-2(t-t_d)}\sin(2\pi(t-t_d)) \end{bmatrix}.$$

The maximum disturbance peak is approximately 40% of the torque value at $t = 2.0$ s. In all graphics presented in the remainder of this book, the dashed line represents joint 1, the solid line represents joint 2 and the dotted line represents joint 3.

$\mathcal{H}_\infty$ control: For the $\mathcal{H}_\infty$ controller designed via game theory, described in Sect. 3.3, the attenuation level found according to Eq. 3.16 is $\gamma = 2.2$. The weighting matrices used are:

$$Q_1 = 2I_3, \quad Q_2 = I_3, \quad Q_{12} = 0, \quad \text{and} \quad R = 3.5I_3.$$

Mixed $\mathcal{H}_2/\mathcal{H}_\infty$ control: The experiment for this controller is performed based on the following weighting matrices:

$$R = \begin{bmatrix} 4.14 & 0 & 0 \\ 0 & 4 & 0 \\ 0 & 0 & 3.86 \end{bmatrix}, \quad Q_1 = \begin{bmatrix} 0.15 & 0 & 0 & 0 & 0 & 0 \\ 0 & 0.2 & 0 & 0 & 0 & 0 \\ 0 & 0 & 0.25 & 0 & 0 & 0 \\ 0 & 0 & 0 & 0.6 & 0 & 0 \\ 0 & 0 & 0 & 0 & 0.8 & 0 \\ 0 & 0 & 0 & 0 & 0 & 1 \end{bmatrix},$$

and $Q_2 = \begin{bmatrix} I_3 & 0 \\ 0 & 4I_3 \end{bmatrix}$. The level of attenuation γ is determined from Eq. 3.26. The experimental joint positions and applied torques are shown in Figs. 3.3, 3.4, 3.5 and 3.6. Table 3.1 shows the values of the cost functions J_1 and J_2, see Eqs. 3.19 and 3.20, and level of attenuation γ for the $\mathcal{H}_\infty$ and mixed $\mathcal{H}_2/\mathcal{H}_\infty$ controllers. Note that we consider also an $\mathcal{H}_2$ controller which is easily obtained considering $\gamma \to \infty$ in the $\mathcal{H}_\infty$ controller. We observe that the disturbance rejection constraint, $J_1 \leq 0$, is satisfied for both $\mathcal{H}_\infty$ and mixed $\mathcal{H}_2/\mathcal{H}_\infty$ controllers. The greater $\mathcal{H}_\infty$ controller capability to attenuate disturbances, represented by a lower value of γ, is confirmed by the lower value of the performance index J_1.

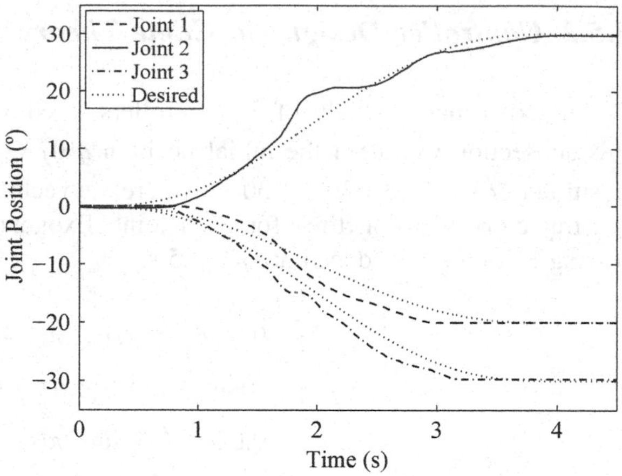

Fig. 3.3 Joint positions, $\mathcal{H}_\infty$ control designed via game theory

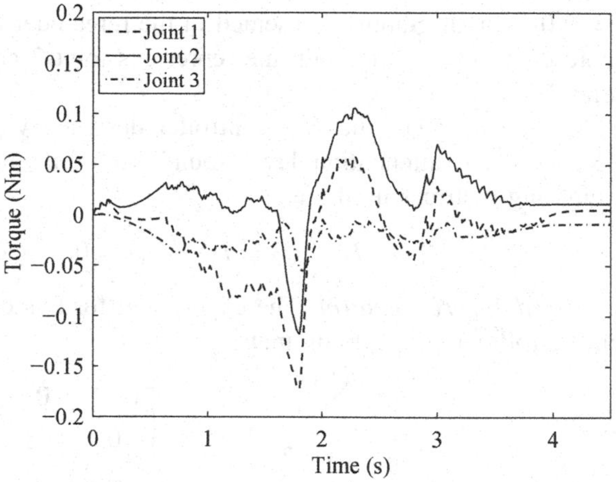

Fig. 3.4 Applied torques, $\mathcal{H}_\infty$ control designed via game theory

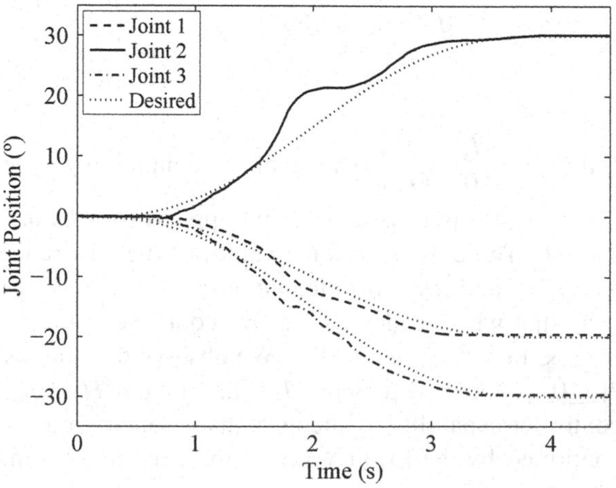

Fig. 3.5 Joint positions, nonlinear mixed $\mathcal{H}_2/\mathcal{H}_\infty$ control

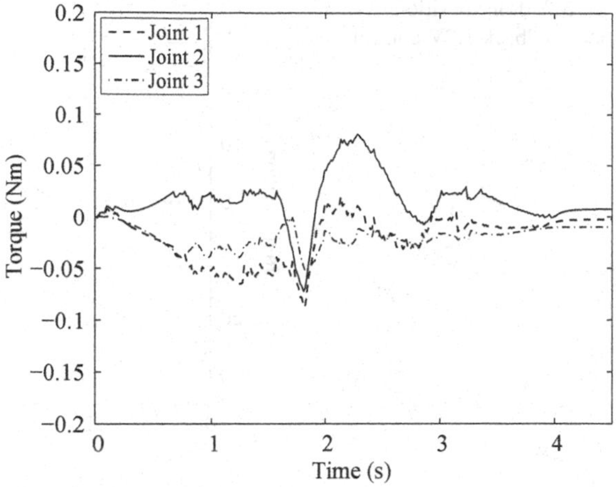

Fig. 3.6 Applied torques, nonlinear mixed $\mathcal{H}_2/\mathcal{H}_\infty$ control

Table 3.1 Cost functions, J_1 and J_2, and attenuation level, γ

	$\mathcal{H}_2$	$\mathcal{H}_\infty$	$\mathcal{H}_2/\mathcal{H}_\infty$
J_1	–	−1.87	−0.563
J_2	0.027	–	0.087
γ	–	2.2	3.0

3.5.3 Controller Design via Linear Matrix Inequalities

State Feedback Control: To apply the control algorithm described in Sect. 3.4, the manipulator should be represented by Eq. 3.29. The parameters $\rho(\tilde{x})$ chosen are the state representing the position errors of joints 2 and 3, i.e.,

$$\rho(\tilde{x}) = [\tilde{q}_2 \;\; \tilde{q}_3]^T.$$

This choice is based on the fact that the inertia matrix $M(q)$ and the Coriolis matrix $C(q, \dot{q})$ are functions of the positions of joints 2 and 3. Consequently, they are functions of the position errors. The system outputs, z_1 and z_2, are the position and velocity errors and the control variable, u, respectively. Hence, the system can be described by:

$$\begin{aligned} A(\rho(x)) &= A(\rho(\tilde{x})), \\ B_1(\rho(x)) &= B, \\ B_2(\rho(x)) &= B, \\ C_1(\rho(x)) &= I_6, \\ C_2(\rho(x)) &= 0, \end{aligned}$$

where the matrices $A(\rho(\tilde{x}))$ and B are defined in Eq. 3.4. The compact set P is defined as $\rho(\tilde{x}) \in [-30, 30]^\circ \times [-30, 30]^\circ$. The parameter variation rate is bounded by $|\dot{\rho}| \leq 50^\circ/\text{s}$. The basis functions selected are

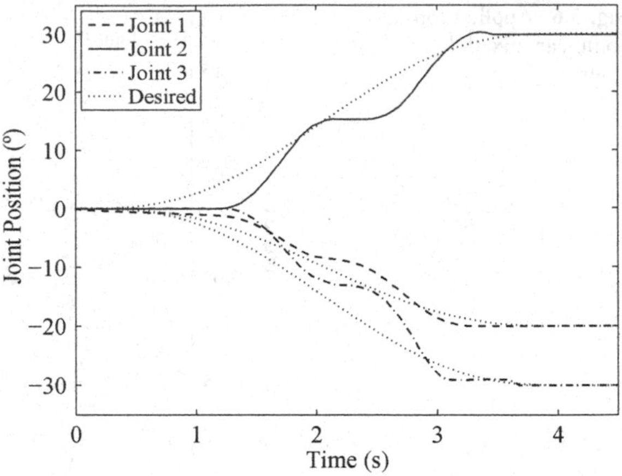

Fig. 3.7 Joint positions, state feedback LPV control

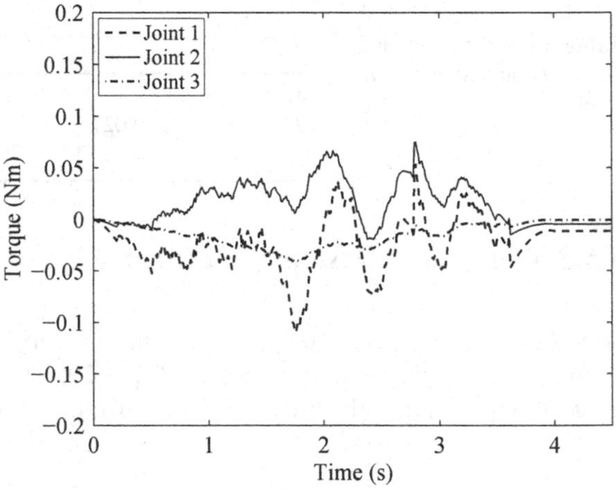

Fig. 3.8 Applied torques, state feedback LPV control

$$\phi_1(\rho(\tilde{x})) = 1,$$
$$\phi_2(\rho(\tilde{x})) = \cos(\tilde{q}_2),$$
$$\phi_3(\rho(\tilde{x})) = \cos(\tilde{q}_3).$$

The matrix $X(\rho(\tilde{x}))$, when represented in this basis is given by:

$$X(\rho(\tilde{x})) = \phi_1(\rho(\tilde{x}))X_1 + \phi_2(\rho(\tilde{x}))X_2 + \phi_3(\rho(\tilde{x}))X_3.$$

The parameter space is divided in $L = 5$ grid points in each dimension, which means that 125 LMIs have to be solved for the X_i variables. The best attenuation level found is $\gamma = 1.2$. Experimental results (joint positions and applied torques) are shown in Figs. 3.7 and 3.8.

Output Feedback Control: The selected parameters for this controller, which are part of the state vector are:

$$\rho(\tilde{x}) = [\tilde{q}_2 \;\; \tilde{q}_3]^T.$$

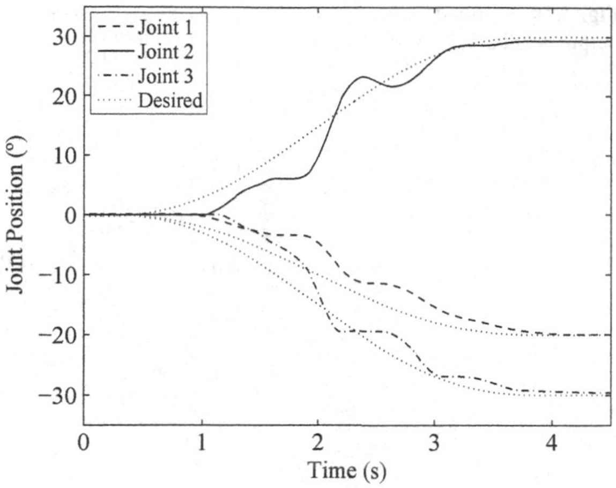

Fig. 3.9 Joint positions, output feedback LPV control

The parameter space P is defined as $\rho(\tilde{x}) \in [30\;\; -30]^\circ \times [30\;\; -30]^\circ$. The parameter variation rate is bounded by $|\dot{\rho}| \leq 90^\circ/\text{s}$. The unknown matrices $X(\rho(\tilde{x}))$ and $Y(\rho(\tilde{x}))$ were defined as follows:

$$X(\rho(\tilde{x})) := \phi_1(\rho(\tilde{x}))X_1,$$
$$Y(\rho(\tilde{x})) := \psi_1(\rho(\tilde{x}))Y_1 + \psi_2(\rho(\tilde{x}))Y_2 + \psi_3(\rho(\tilde{x}))Y_3,$$

where $\phi_1(\rho(\tilde{x})) = 1$, $\psi_1(\rho(\tilde{x})) = 1$, $\psi_2(\rho(\tilde{x})) = \sin(q_2) + \cos(q_2)$, and $\psi_3(\rho(\tilde{x})) = \sin(q_3) + \cos(q_3)$.

For the algorithm described in Eqs. 3.35–3.42, the matrices $\widehat{A}_k$, $\widehat{B}_k$, $\widehat{C}_k$, and D_k are assumed to be constant. They do not depend on the basis functions, like X does, when $\phi_1(\rho(\tilde{x}))$ is set to 1. The parameter space was divided in $L = 5$ grid points for each parameter. The best levels of attenuation found was $\gamma = 2.359$. Experimental results (joint positions and applied torques) are shown in Figs. 3.9 and 3.10.

We compare the performance of the four nonlinear $\mathcal{H}_\infty$ controllers presented in this section using the performance indexes defined in Eqs. 2.23 and 2.24. A total of five experiments were performed for each controller to compute a mean value for the $\mathcal{L}_2$ norm and the sum of the applied torques $E[\tau]$. The experimental results shown in Figs. 3.3, 3.4, 3.5, 3.6, 3.7, 3.8, 3.9 and 3.10 correspond to the samples that are closest to the mean values of $\mathcal{L}_2(\tilde{x})$ and $E[\tau]$, shown in Table 3.2.

Note that the nonlinear $\mathcal{H}_\infty$ controllers based on the game theory presents the best $\mathcal{L}_2[\tilde{x}]$ norm value, which can be confirmed by the best desired trajectory tracking shown in the graphics. Also, the $\mathcal{H}_2/\mathcal{H}_\infty$ controller presented the lowest energy consumption, which is explained by the $\mathcal{H}_2$ component of this functional.

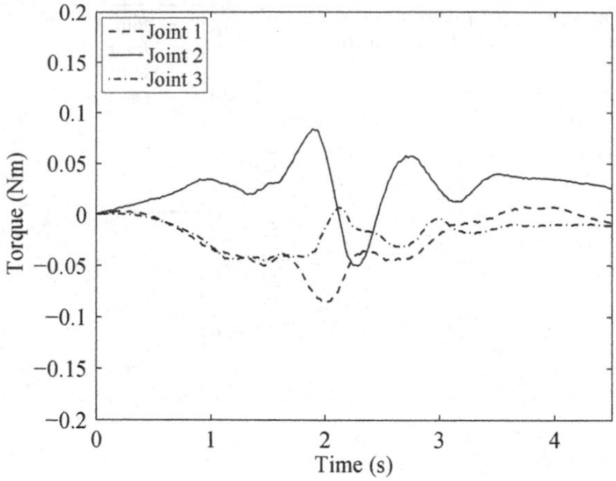

Fig. 3.10 Applied torques, output feedback LPV control

Table 3.2 Performance indexes

Controller	$\mathcal{L}_2[\tilde{x}]$	$E[\tau]$ (N m s)
$\mathcal{H}_\infty$ control through game theory	0.0149	0.3709
Mixed $\mathcal{H}_2/\mathcal{H}_\infty$ control	0.0141	0.2917
State feedback LPV control	0.0271	0.3945
Output feedback LPV control	0.0266	0.3575

References

1. Apkarian P, Adams RJ (1998) Advanced gain-scheduling techniques for uncertain systems. IEEE Trans Control Syst Technol 6(1):21–32
2. Basar T, Bernhard P (1990) $\mathcal{H}_\infty$-Optimal control and related minimax problems. Birkhauser, Berlin
3. Chen BS, Chang YC (1997) Nonlinear mixed $\mathcal{H}_2/\mathcal{H}_\infty$ control for robust tracking design of robotic systems. Int J Control 67(6):837–857
4. Chen BS, Lee TS, Feng JH (1994) A nonlinear $\mathcal{H}_\infty$ control design in robotic systems under parameter perturbation and external disturbance. Int J Control 59(2):439–461
5. Gahinet P, Apkarian P (1994) A linear matrix inequality approach to $\mathcal{H}_\infty$ control. Int J Robust Nonlinear Control 4(4):421–448
6. Gahinet P, Nemiroviski A, Laub AJ, Chilali M (1995) LMI control toolbox. The MathWorks, Inc., Natick
7. Johansson R (1990) Quadratic optimization of motion coordination and control. IEEE Trans Autom Control 35(11):1197–1208
8. Postlethwaite I, Bartoszewicz A (1998) Application of non-linear $\mathcal{H}_\infty$ control to the tetrabot robot manipulator. Proc Inst Mech Eng: Part I. J Syst Control Eng 212(16):459–465
9. Wu F, Yang XH, Packard A, Becker G (1996) Induced $\mathcal{L}_2$-norm control for LPV systems with bounded parameter variation rates. Int J Robust Nonlinear Control 6(9–10):983–998

Chapter 4
Adaptive Nonlinear $\mathcal{H}_\infty$ Control

4.1 Introduction

Generally speaking, three classes of control strategies are available in the literature for robotic manipulators. They are categorized according to the level of knowledge the designer has about the dynamic model of the robot (see for instance [1–11], and references therein). Strategies in the first class consider that both the mathematical model and the values of the kinematic and dynamic parameters are well known and available to the controller. Parametric uncertainties are treated as perturbations acting on the system to be suppressed by a robust controller. The controllers presented so far in this book belong in this class.

The second class of control strategies considers that the analytical expressions of the dynamic model of the manipulator are known, but the parameters values used in the controller design are imprecise. In this case, an adaptive control law can be used to estimate the uncertain parameters. The linear parameterization property has been extensively used to deal with this problem [6, 11]; it states that the dynamic model of robotic manipulators can be expressed as the product of a signal-dependent matrix, namely the regression matrix, and a parameter-dependent vector which contains the uncertain parameters. These parameters are updated online by an error-based control law.

Controllers in the third class consider that both the dynamic model of the manipulator and its parameters' values are unknown. In this case, control strategies such as neural networks, fuzzy logic, and genetic algorithms, have been used to estimate the dynamic model of the manipulator, see for instance [2, 3].

Adaptive nonlinear $\mathcal{H}_\infty$ controllers for robot manipulators are proposed in [2–4]; they fall within both the second and third classes of controllers aforementioned. A model-based adaptive algorithm is proposed in [4], where a robust tracking design considers that the unknown parameters can be learned by classical adaptive update laws. In [3], an adaptive neural-network tracking control with a guaranteed $\mathcal{H}_\infty$ performance is developed. The neural network is employed to approximate the robot dynamics and the torque disturbances; the $\mathcal{H}_\infty$ controller is

A. A. G. Siqueira et al., *Robust Control of Robots*,
DOI: 10.1007/978-0-85729-898-0_4,

designed to attenuate the effect of the approximation (generated by the neural network algorithm and considered as the disturbance to be attenuated) on the tracking error.

In this chapter, we assume that the nominal model of the robot manipulator is known and we estimate only the uncertain part of the robot dynamics through approaches based on linear parameterization and neural networks [7].

This chapter is organized as follows: Sect. 4.2 describes the adaptive control strategy based on the linear parameterization property, with an $\mathcal{H}_\infty$ attenuation. Section 4.3 presents the set of neural networks used to estimate the robot dynamics and the solution for the adaptive neural network-based control problem. Section 4.4 presents the results of applying the proposed controllers to the UARM robot.

4.2 Adaptive $\mathcal{H}_\infty$ Control

One of the most widely used adaptive control strategies for robot manipulators takes advantage of the robot's dynamic model linear parameterization property:

Property 4.1 Consider the dynamic matrices of a robot manipulator, $M(q)$, $C(q,\dot{q})$, $F(\dot{q})$, and $G(q)$ and a given joint trajectory, $q_s \in \Re^n$, with its derivatives, $\dot{q}_s$ and $\ddot{q}_s$. It is always possible to find an $n \times p$-dimensional matrix of known functions, $Y(\ddot{q}_s, \dot{q}_s, q_s)$, and a p-dimensional vector with components depending on manipulator parameters, θ, such that:

$$M(q_s)\ddot{q}_s + C(q_s,\dot{q}_s)\dot{q}_s + F(\dot{q}_s) + G(q_s) = Y(\ddot{q}_s,\dot{q}_s,q_s)\theta. \tag{4.1}$$

For the adaptive nonlinear $\mathcal{H}_\infty$ controllers described in this chapter, the dynamic equation of the state tracking error for a robot manipulator must be written in a suitable form. Consider again the dynamic equation of a robot manipulator with exogenous torque disturbance:

$$\tau + \tau_d = M(q)\ddot{q} + C(q,\dot{q})\dot{q} + F(\dot{q}) + G(q).$$

Taking into account the formulation presented in Sect. 3.3 for the state tracking error (3.3) and the state transformation (3.5), the state space equation is given as:

$$\dot{\tilde{x}} = A_T(\tilde{x},t)\tilde{x} + B_T(\tilde{x},t)u + B_T(\tilde{x},t)w, \tag{4.2}$$

with

$$A_T(\tilde{x},t) = T_0^{-1}\begin{bmatrix} -T_{12}^{-1}T_{11} & T_{12}^{-1} \\ 0 & -M_0^{-1}(q)C_0(q,\dot{q}) \end{bmatrix} T_0,$$

$$B_T(\tilde{x},t) = T_0^{-1}\begin{bmatrix} 0 \\ M_0^{-1}(q) \end{bmatrix},$$

$$u = T_{12}(-f(x_e) + \tau),$$
$$w = M(q)T_{12}M^{-1}(q)\tau_d,$$
$$f(x_e) = M(q)(\ddot{q}^d - T_{12}^{-1}T_{11}\dot{\tilde{q}}) + C(q,\dot{q})(\dot{q}^d - T_{12}^{-1}T_{11}\tilde{q}) + F(\dot{q}) + G(q),$$

and $x_e = \left[(\ddot{q}^d)^T(\dot{q}^d)^T(q^d)^T\dot{q}^Tq^T\right]^T$. The resulting applied torques are given by $\tau = f(x_e) + T_{12}^{-1}u$. These torques are essentially the same presented in Eq. 3.7 for the nonlinear $\mathcal{H}_\infty$ control. The difference is that, here, we use the complete dynamic matrices ($M(q)$, $C(q,\dot{q})$, $F(\dot{q})$, and $G(q)$) instead of the nominal dynamic matrices ($M_0(q)$, $C_0(q,\dot{q})$, $F_0(\dot{q})$, and $G_0(q)$). We can consider the model-based term $f(x_e)$ as composed of nominal and uncertain terms defined by:

$$f(x_e) = f_0(x_e) + \Delta f(x_e), \tag{4.3}$$

where

$$f_0(x_e) = M_0(q)(\ddot{q}^d - T_{12}^{-1}T_{11}\dot{\tilde{q}}) + C_0(q,\dot{q})(\dot{q}^d - T_{12}^{-1}T_{11}\tilde{q}) + F_0(\dot{q}) + G_0(q),$$
$$\Delta f(x_e) = \Delta M(q)(\ddot{q}^d - T_{12}^{-1}T_{11}\dot{\tilde{q}}) + \Delta C(q,\dot{q})(\dot{q}^d - T_{12}^{-1}T_{11}\tilde{q}) + \Delta F(\dot{q}) + \Delta G(q).$$

The controller design presented in Sect. 3.3 considers the uncertain terms as part of the disturbance w. Here, the assumption of linear parameterization of $\Delta f(x_e)$ is used to design an adaptive control law to learn the behavior of this uncertain term. According to Property 4.1, $\Delta f(x_e)$ can be expressed as:

$$\Delta f(x_e) = Y(\ddot{q}^d, \dot{q}^d, q^d, \dot{q}, q)\theta. \tag{4.4}$$

The adaptive controller described aims to estimate the uncertain part $\Delta f(x_e)$ and to satisfy a desired $\mathcal{H}_\infty$ tracking performance. Considering the above formulation, we are able to state the

Adaptive Nonlinear $\mathcal{H}_\infty$ Control Problem: Given a level of attenuation γ, find an adaptive state feedback controller:

$$\dot{\hat{\theta}} = \alpha(t,\tilde{x}), \quad \tau = f_0(x_e) + Y\hat{\theta} + T_{12}^{-1}u,$$

such that the closed-loop system satisfies the following performance index

$$\int_0^T (\tilde{x}^TQ\tilde{x} + u^TRu)dt \le \tilde{x}^T(0)P_0\tilde{x}(0) + \tilde{\theta}^T(0)S_0\tilde{\theta}(0) + \gamma^2\int_0^T (w^Tw)dt,$$

for some matrices $Q = Q^T > 0$, $R = R^T > 0$, $P_0 = P_0^T > 0$, and $S_0 = S_0^T > 0$, where $\widetilde{\theta} = \theta - \hat{\theta}$ denotes the parameter estimation error and $\hat{\theta}$ is the estimated parameter vector.

The solution for this nonlinear control problem can be found following the game theory-based derivation presented in Chap. 3 and the adaptive controller proposed in [4]. For this purpose, consider the following Lyapunov function:

$$V(\widetilde{x}, t) = \frac{1}{2}\widetilde{x}^T P(\widetilde{x}, t)\widetilde{x} + \frac{1}{2}\widetilde{\theta}^T S\widetilde{\theta},$$

where $P(\widetilde{x}, t)$ is the positive definite symmetric solution of the Riccati equation 3.11. The adaptive control law

$$\begin{aligned} \dot{\hat{\theta}} &= -S^{-1}Y^T T_{12} B_T^T(\widetilde{x}, t) P(\widetilde{x}, t)\widetilde{x}, \\ \tau &= f_0(x_e) + Y\hat{\theta} - T_{12}^{-1} R^{-1} B_T^T(\widetilde{x}, t) P(\widetilde{x}, t)\widetilde{x}, \end{aligned} \tag{4.5}$$

is a solution for the adaptive nonlinear $\mathcal{H}_\infty$ control problem, for any positive definite symmetric matrix S. Analogously to Sect. 3.3 and based on the solution of the algebraic equation (3.13), we can find the following

Simplified solution:

$$\begin{aligned} \dot{\hat{\theta}} &= -S^{-1}Y^T T_{12} B^T T_0 \widetilde{x}, \\ \tau &= f_0(x_e) + Y\hat{\theta} - T_{12}^{-1} R^{-1} B^T T_0 \widetilde{x}. \end{aligned} \tag{4.6}$$

In order to guarantee that the estimated parameter vector $\hat{\theta}(t)$ is inside a given constraint region for all t, projection algorithms must be used in this control approach. Consider a pre-assigned constraint region $\Omega_{\hat{\theta}}$ of the parameter $\hat{\theta}$, with $\Omega_{\hat{\theta}} = \{\hat{\theta} : \hat{\theta}^T\hat{\theta} \le M_{\hat{\theta}} + \delta\}$ for some $M_{\hat{\theta}} > 0$ and $\delta > 0$. A projection algorithm can be given as [2]:

$$Proj[\Phi] = \begin{cases} \Phi, & \text{if } \hat{\theta}^T\hat{\theta} \le M_{\hat{\theta}} \text{ or} \\ & \hat{\theta}^T\hat{\theta} > M_{\hat{\theta}} \text{ and } \hat{\theta}^T\Phi \le 0, \\ \Phi - \frac{(\hat{\theta}^T\hat{\theta} - M_{\hat{\theta}})\hat{\theta}^T\Phi}{\delta\hat{\theta}^T\hat{\theta}}\hat{\theta}, & \text{otherwise,} \end{cases} \tag{4.7}$$

where $\Phi = -S^{-1}Y^T T_{12} B^T T_0 \widetilde{x}$.

4.3 Neural Network-Based $\mathcal{H}_\infty$ Control

When the mathematical model of the robotic manipulator is only partially known—for example, when friction components cannot be fully modeled—neural networks can be used to estimate the unknown elements of the model. One advantage of neural network-based strategies is that they do not need information on the system analytical model. In this section, a neural network $\Delta f(x_e, \Theta)$, where Θ is a vector containing the tunable network parameters, is used to approximate the uncertain term $\Delta f(x_e)$ in (4.3). This is an alternative approach to the linear parameterization presented earlier.

We define n neural networks $\Delta f_k(x_e, \theta_k)$, $k = 1, \ldots, n$ composed of nonlinear neurons in every hidden layer and linear neurons in the input and output layers, with adjustable parameters θ_k in the output layers [2, 3]. The single-output neural networks are of the form:

$$\Delta f_k(x_e, \theta_k) = \sum_{i=1}^{p_k} H\left(\sum_{j=1}^{5n} w_{ij}^k x_{ej} + m_i^k\right)\theta_{ki} = \xi_k^T \theta_k, \tag{4.8}$$

where

$$\xi_k = \begin{bmatrix} \xi_{k1} \\ \vdots \\ \xi_{kp_k} \end{bmatrix} = \begin{bmatrix} H\left(\sum_{j=1}^{5n} w_{1j}^k x_{ej} + m_1^k\right) \\ \vdots \\ H\left(\sum_{j=1}^{5n} w_{p_kj}^k x_{ej} + m_{p_k}^k\right) \end{bmatrix}, \quad \theta_k = \begin{bmatrix} \theta_{k1} \\ \vdots \\ \theta_{kp_k} \end{bmatrix},$$

and p_k is the number of neurons in the hidden layer, the weights w_{ij}^k and the biases m_i^k for $1 \leq i \leq p_k$, $1 \leq j \leq 5n$ and $1 \leq k \leq n$ are assumed to be constant and specified by the designer, and $H(.)$ is the hyperbolic tangent function

$$H(z) = \frac{e^z - e^{-z}}{e^z + e^{-z}}.$$

Note that in Eq. 4.8, $5n$ represents five state variables: position, velocity, and the desired position, velocity and acceleration of n joints. The complete neural network is denoted by:

$$\Delta f(x_e, \Theta) = \begin{bmatrix} \Delta f_1(x_e, \theta_1) \\ \Delta f_2(x_e, \theta_2) \\ \vdots \\ \Delta f_n(x_e, \theta_n) \end{bmatrix} = \begin{bmatrix} \xi_1^T \theta_1 \\ \xi_2^T \theta_2 \\ \vdots \\ \xi_n^T \theta_n \end{bmatrix},$$

$$= \begin{bmatrix} \xi_1^T & 0 & \ldots & 0 \\ 0 & \xi_2^T & \vdots & 0 \\ \vdots & \vdots & \ddots & \vdots \\ 0 & 0 & \ldots & \xi_n^T \end{bmatrix} \begin{bmatrix} \theta_1 \\ \theta_2 \\ \vdots \\ \theta_n \end{bmatrix} = \Xi \Theta. \tag{4.9}$$

We assume that there exists a parameter value $\Theta^{\star} \in \Omega_\Theta$ such that $\Delta f(x_e, \Theta^{\star})$ can approximate $\Delta f(x_e)$ as close as possible, where Ω_Θ is a pre-assigned constraint region. When neural networks are used, one possible way of achieving $\mathcal{H}_\infty$ performance is to consider the approximation error $\delta f(x_e) = \Delta f(x_e, \Theta^{\star}) - \Delta f(x_e)$ as a disturbance. This approach was used in [3], where the complete robot dynamic term $f(x_e)$ is approximated by a neural network. Since the the approximation error includes the effects of the uncertain dynamics, however, the necessary property that the disturbance be square-integrable is not simple to be demonstrated.

An alternative procedure is to consider that the approximation error is bounded by a state-dependent function, that is, there exists a function $k(x_e) > 0$ such that $|\delta f(x_e)_i| \leq k(x_e)$, for all $1 \leq i \leq n$. With this weaker assumption, it is possible to include a variable structure control (VSC) into the control strategy and the disturbance remains square-integrable, since it is composed only of the external torque disturbances.

With these assumptions, the adaptive control problem with $\mathcal{H}_\infty$ performance for robotic manipulators can be reformulated as follows.

Adaptive Neural Network Nonlinear $\mathcal{H}_\infty$ Control Problem: Given a level of attenuation γ, find an adaptive neural network state feedback controller

$$\dot{\Theta} = \beta(t, \widetilde{x}),$$
$$\tau = f_0(x_e) + \Xi\Theta + T_{12}^{-1}u + T_{12}^{-1}k(x_e)u_s,$$

such that the following performance index is achieved:

$$\int_0^T (\widetilde{x}^T Q\widetilde{x} + u^T Ru)dt \leq \widetilde{x}^T(0)P_0\widetilde{x}(0) + \widetilde{\Theta}^T(0)Z_0\widetilde{\Theta}(0) + \gamma^2 \int_0^T (w^T w)dt,$$

for some matrices $Q = Q^T > 0, R = R^T > 0, P_0 = P_0^T > 0,$ and $Z_0 = Z_0^T > 0$, where $\widetilde{\Theta} = \Theta - \Theta^*$ denotes the neural parameter estimation error, and u_s is the VSC control used to eliminate the effect of the approximation error. u and u_s are defined in the following.

Considering the results of the previous section and the Lyapunov function

$$V(\widetilde{x}, t) = \frac{1}{2}\widetilde{x}^T P(\widetilde{x}, t)\widetilde{x} + \frac{1}{2}\widetilde{\theta}^T Z\widetilde{\theta},$$

where $P(\widetilde{x},t)$ is the solution of the Riccati equation 3.11, the adaptive neural network control law

$$\begin{aligned}\dot{\Theta} &= -Z^{-1}\Xi^T T_{12} B_T^T(\widetilde{x},t)P(\widetilde{x},t)\widetilde{x},\\ \tau &= f_0(x_e) + \Xi\Theta - T_{12}^{-1}R^{-1}B_T^T(\widetilde{x},t)P(\widetilde{x},t)\widetilde{x}\\ &\quad - T_{12}^{-1}k(x_e)sgn(B_T^T(\widetilde{x},t)P(\widetilde{x},t)\widetilde{x}),\end{aligned}$$

is a solution for the adaptive neural network nonlinear $\mathcal{H}_\infty$ control problem, for any positive definite symmetric matrix Z. Note that the control inputs u and u_s result in:

$$\begin{aligned}u &= -R^{-1}B_T^T(\widetilde{x},t)P(\widetilde{x},t)\widetilde{x},\\ u_s &= -sgn(B_T^T(\widetilde{x},t)P(\widetilde{x},t)\widetilde{x}).\end{aligned}$$

Again, we can compute a

Simplified solution:

$$\begin{aligned}\dot{\Theta} &= -Z^{-1}\Xi^T T_{12} B^T T_0\widetilde{x},\\ \tau &= f_0(x_e) + \Xi\Theta - T_{12}^{-1}R^{-1}B^T T_0\widetilde{x} - T_{12}^{-1}k(x_e)sgn(B^T T_0\widetilde{x}),\end{aligned} \tag{4.10}$$

where T_0 is a solution of Eq. 3.13. The projection algorithm described in Sect. 4.2 (Eq. 4.7) is also used in order to guarantee that the parameter Θ is constrained to the compact set Ω_Θ.

As cited in [3, p. 17], neural network systems may yield a poor approximation of a nonlinear continuous function if there exist some discontinuous uncertainties in the function, causing instability in the control system. It was emphasized in that reference that the $\mathcal{H}_\infty$ attenuation property solves this problem. The same stability property is guaranteed in the procedure proposed in this section, mainly because the neural network is used only to ameliorate the performance of the mathematical model.

4.4 Examples

In this section we present practical guidelines to implement the adaptive control strategies described in this chapter. These guidelines show how to compute the regression matrix and the adaptive parameter vector for the model-based methodology (Sect. 4.2), and the neural networks for the network-based one (Sect. 4.3). The proposed controllers can be designed by the user through the Control Environment for Robots (CERob). We present experimental results obtained with the UARM manipulator.

4.4.1 Design Procedures

The most important issues related to the design of model-based adaptive control strategies are the definition of the estimation parameter vector θ and the computation of the regression matrix. As a starting point, the estimation parameters can be selected as the manipulators dynamic parameters or a (possibly nonlinear) combination of them. For example, the linear parameterization of the three-link robot manipulator UARM used to obtain the results presented in Sect. 4.4.2 is:

$$\begin{array}{lll}\theta_1=\Delta(m_1l_{c1}^2), & \theta_2=\Delta(m_2l_1^2), & \theta_3=\Delta(m_2l_1l_{c2}),\\ \theta_4=\Delta(m_3l_1^2), & \theta_5=\Delta(m_3l_{c3}^2), & \theta_6=\Delta(m_3l_1l_{c3}),\\ \theta_7=\Delta I_1, & \theta_8=\Delta I_3, & \theta_9=\Delta f_1,\\ \theta_{10}=\Delta f_2, & \theta_{11}=\Delta f_3. & \end{array}$$

With this selection, the regression matrix $Y(\ddot{q}^d,\dot{q}^d,q^d,\dot{q},q)$ is given by:

$$Y(\cdot)=\begin{bmatrix} Y_{11}(\cdot) & Y_{12}(\cdot) & \cdots & Y_{1,11}(\cdot)\\ Y_{21}(\cdot) & Y_{22}(\cdot) & \cdots & Y_{2,11}(\cdot)\\ Y_{31}(\cdot) & Y_{32}(\cdot) & \cdots & Y_{3,11}(\cdot)\end{bmatrix},$$

where

$$\begin{aligned}
&Y_{11}(\cdot)=2y_{11}+y_{12},\ Y_{12}(\cdot)=y_{11},\\
&Y_{13}(\cdot)=2\cos(q_2)y_{11}+\cos(q_2)y_{12}-\sin(q_2)\dot{q}_2y_{21}-\sin(q_2)(\dot{q}_1+\dot{q}_2)y_{22},\\
&Y_{14}(\cdot)=(2+2\cos(q_2))y_{11}+(1+\cos(q_2))y_{12}-\sin(q_2)\dot{q}_2y_{21}-\sin(q_2)(\dot{q}_1+\dot{q}_2)y_{22},\\
&Y_{15}(\cdot)=y_{11}+y_{12}+y_{13},\\
&Y_{16}(\cdot)=(2\cos(q_2+q_3)+2\cos(q_3))y_{11}+(\cos(q_2+q_3)+2\cos(q_3))y_{12}\\
&\qquad+(\cos(q_2+q_3)+\cos(q_3))y_{13}-\sin(q_2+q_3)\dot{q}_2y_{21}\\
&\qquad-(\sin(q_2+q_3)+\sin(q_3))\dot{q}_3y_{21}-\sin(q_2+q_3)(\dot{q}_1+\dot{q}_2)y_{22}\\
&\qquad-(\sin(q_2+q_3)+\sin(q_3))\dot{q}_3y_{22}-(\sin(q_2+q_3)+\sin(q_3))(\dot{q}_1+\dot{q}_2+\dot{q}_3)y_{23},\\
&Y_{17}(\cdot)=2y_{11}+y_{12},\ Y_{18}(\cdot)=y_{11}+y_{12}+y_{13},\\
&Y_{19}(\cdot)=\dot{q}_1,\ Y_{1,10}(\cdot)=0,\ Y_{1,11}(\cdot)=0,\\
&Y_{21}(\cdot)=2y_{11}+y_{12},\ Y_{22}(\cdot)=0,\\
&Y_{23}(\cdot)=\cos(q_2)y_{11}+\sin(q_2)\dot{q}_1y_{21},\\
&Y_{24}(\cdot)=(1+\cos(q_2))y_{11}+y_{12}+\sin(q_2)\dot{q}_1y_{21},\\
&Y_{25}(\cdot)=y_{11}+y_{12}+y_{13},\\
&Y_{26}(\cdot)=(\cos(q_2+q_3)+2\cos(q_3))y_{11}+2\cos(q_3)y_{12}+\cos(q_3)y_{13}+\sin(q_2+q_3)\dot{q}_1y_{21}\\
&\qquad-\sin(q_3)\dot{q}_3y_{21}-\sin(q_3)\dot{q}_3y_{22}-\sin(q_3)(\dot{q}_1+\dot{q}_2+\dot{q}_3)y_{22},\\
&Y_{27}(\cdot)=y_{11}+y_{12},\ Y_{28}(\cdot)=y_{11}+y_{12}+y_{13},\\
&Y_{29}(\cdot)=0,\ Y_{2,10}(\cdot)=\dot{q}_2,\ Y_{2,11}(\cdot)=0,\\
&Y_{31}(\cdot)=0,\ Y_{32}(\cdot)=0,\ Y_{33}(\cdot)=0,\ Y_{34}(\cdot)=0,\\
&Y_{35}(\cdot)=y_{11}+y_{12}+y_{13},
\end{aligned}$$

$$
\begin{aligned}
Y_{36}(\cdot) =& \cos(q_2+q_3)y_{11}+\cos(q_3)y_{11}+\cos(q_3)y_{12}+\sin(q_2+q_3)\dot{q}_1 y_{21}+\sin(q_3)\dot{q}_1 y_{21} \\
& -\sin(q_3)\dot{q}_2 y_{21}+\sin(q_3)(\dot{q}_1+\dot{q}_2)y_{22}, \\
Y_{37}(\cdot) =& 0,\ Y_{38}(\cdot)=y_{11}+y_{12}+y_{13}, \\
Y_{39}(\cdot) =& 0,\ Y_{3,10}(\cdot)=0,\ Y_{3,11}(\cdot)=\dot{q}_3,
\end{aligned}
$$

and

$$
\begin{aligned}
y_{11} &= (\ddot{q}_1^d - T_{12}^{-1}T_{11}\dot{\tilde{q}}_1), & y_{12} &= (\ddot{q}_2^d - T_{12}^{-1}T_{11}\dot{\tilde{q}}_2), \\
y_{13} &= (\ddot{q}_3^d - T_{12}^{-1}T_{11}\dot{\tilde{q}}_3), & y_{21} &= (\dot{q}_1^d - T_{12}^{-1}T_{11}\tilde{q}_1), \\
y_{22} &= (\dot{q}_2^d - T_{12}^{-1}T_{11}\tilde{q}_2), & y_{23} &= (\dot{q}_3^d - T_{12}^{-1}T_{11}\tilde{q}_3).
\end{aligned}
$$

The MATLAB® code used to compute the above function can be found in the file `CERob\Underactuated\pr31_adap.m`, along with all other files mentioned in this section.

The implementation of the adaptive nonlinear $\mathcal{H}_\infty$ controller follows the expressions in Eq. 4.6, with the inclusion of the projection algorithm. This algorithm prevents the estimation parameter vector to increase without bounds. In the following MATLAB® code, the constraint region is defined by `Mtheta` and `delta`, respectively, $M_{\hat{\theta}}$ and δ in (4.7).

File: uarm_cont_ADP.m

```
...

%compute the composite errors
y1 = qdd_d(:,i)  - inv(T12)*T11*errorqd;
y2 = qd_d(:,i)  - inv(T12)*T11*errorq;

%compute the Regression Matrix
 Y = pr31_adap(theta,omega,y1,y2);

%implement the Projection Algorithm
FI(:,i)  = -  inv(S)*Y'*T12*Be'*T0*[errorq;errorqd];

if i > 1
   ThetaT =  Theta(:,i-1)'*Theta(:,i-1);
   if (ThetaT <= Mtheta | ...
    (ThetaT > Mtheta & Theta(:,i-1)'*FI(:,i) <=0))
      Theta_d(:,i) = FI(:,i);
   else
      Theta_d(:,i)=FI(:,i) -(((ThetaT-Ka)*Theta(:,i-1)'*...
        FI(:,i))/(delta*ThetaT))*Theta(:,i-1);
   end;
end
```

```
%update the adapted parameters
Theta(:,i) = Theta(:,i-1) + Theta_d(:,i)*dt;

%compute the nominal term
Fxe_o(:,i) = Mest*y1 + Cest*y2+ Fest*qd(:,i) + Gest;

%compute the applied torques
u(:,i) = -inv(R)*Be'*T0*[errorq;errorqd];

tau(:,i) = Fxe_o(:,i) + inv(T12)*u(:,i) + Y*Theta(:,i);
```

The controller gains `T11`, `T12`, and `T0` used in this code are computed according to Sect. 3.3 through the selection of the weighting matrices Q and R. These matrices and the adaptive gain S, along with the constraint region parameters, can be added by selecting `Adaptive Hinf` in the menu `Controller` and pressing the *Controller Design* button (see Fig. 4.1).

To implement the adaptive neural network-based controller, the main difference resides in the computation of matrix Ξ, used in the adaptation law. This matrix is given by a set of neural networks, whose inputs are the desired joint accelerations, velocities and positions, and the actual joint velocities and positions. The following MATLAB® code shows how to compute Ξ for a general robot manipulator with n joints and considering p_k neurons in the hidden layer. We show in Sect. 4.4.3 the special case where $n = 3$ and $p_k = 7$.

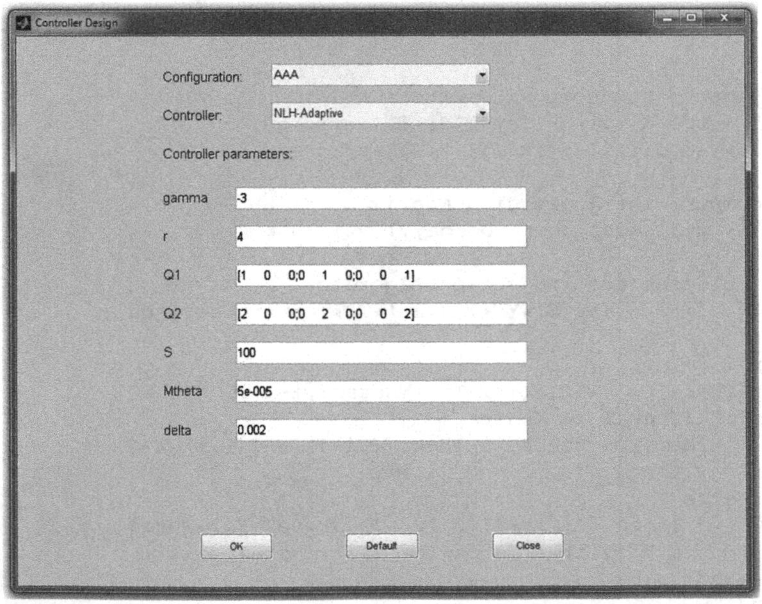

Fig. 4.1 Controller design box for the adaptive nonlinear $\mathcal{H}_\infty$ controller

File: uarm_cont_NET.m

```
...

%compute the input signal for the hidden layer
psi(i) = (sum(q(:,i)) + sum(qd(:,i)) - sum(q_d(:,i)) ...
      - sum(qd_d(:,i)) - sum(qdd_d(:,i)));

%compute the neural network matrix XI


XI_aux=[];
for k=1:n
          for j=1:pk
               x(k,j) = tanh(psi(i)+m(k,j));
          end
          XI_aux = [XI_aux,sprintf('x(%1.0f,:),',k)];
end
XI_aux = XI_aux(1:end-1);
eval(['XI = blkdiag(' XI_aux ')']);
```

The neural network parameters can be selected by choosing `Adaptive NN Hinf` in the menu `Controller` and pressing the *Controller Design* button.

4.4.2 Model-Based Controller

In this section, we present the results obtained with the implementation of the adaptive nonlinear $\mathcal{H}_\infty$ control on the robotic manipulator UARM [10]. The parameter vector, θ, is defined as a combination of the dynamic parameters of the manipulator in order to derive the linear parameterization of $\Delta f(x_e)$. The structure of θ and the related regression matrix $Y(\cdot)$ for the results presented here are those shown in Sect. 4.4.1.

For the experiments presented in this section and the next one, an external disturbance of the following form is added at $t_d = 2s$, where $\mu = 0.72$:

$$\tau_d = \begin{bmatrix} 0.04e^{\frac{-(t-t_d)^2}{2\mu^2}}\sin(3.2\pi t) \\ 0.04e^{\frac{-(t-t_d)^2}{2\mu^2}}\sin(2.4\pi t) \\ 0.02e^{\frac{-(t-t_d)^2}{2\mu^2}}\sin(1.6\pi t) \end{bmatrix}.$$

The robot motion is performed considering an initial position $q(0) = [0^\circ\ 0^\circ\ 0^\circ]^T$ and desired final position $q(T_f) = [-20^\circ \quad 30^\circ \quad -30^\circ]^T$, where the vector $T_f = [4.0\ 4.0\ 4.0]$ s contains the trajectory duration time for each joint. Figures 4.2 and 4.3 show the experimental results for the adaptive nonlinear $\mathcal{H}_\infty$ control, with $\theta(0) = [0\ldots 0]^T_{11\times 1}$, $S = 100$, $\gamma = 2$, and weighting matrices $R = 3.6I_3$ and

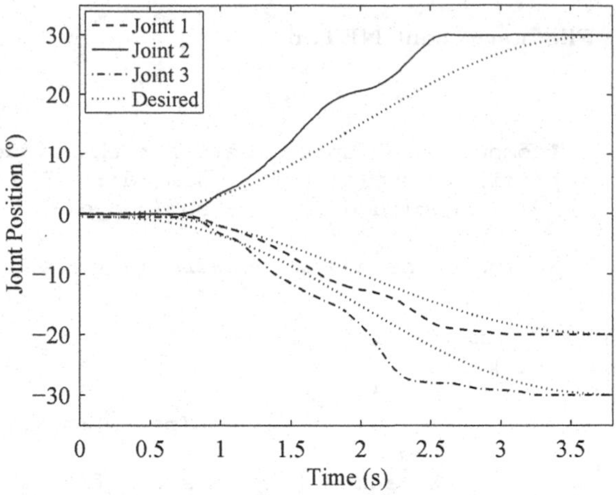

Fig. 4.2 Joint positions, adaptive nonlinear $\mathcal{H}_\infty$ control

$$Q = \begin{bmatrix} 30I_3 & 0 \\ 0 & 30I_3 \end{bmatrix}.$$

4.4.3 Neural Network-Based Controller

To implement the adaptive nonlinear $\mathcal{H}_\infty$ control via neural networks in the three-link robot manipulator UARM, three neural networks ($\Delta f_k(x_e, \theta_k)$, $k = 1, \ldots, 3$) must be computed. The inputs of the neural networks are the elements of x_e, i.e., the desired joint accelerations, velocities and positions, and the actual joint velocities and positions. Define the following auxiliary variable to deal with the input signals:

$$\psi = \sum_{i=1}^{3} (q_i - q_i^d) + \sum_{i=1}^{3} (\dot{q}_i - \dot{q}_i^d) - \sum_{i=1}^{3} \ddot{q}_i^d. \tag{4.11}$$

This variable is the input for the hyperbolic tangent functions working here as the activation functions of the hidden layer. The weights w_{ij}^k assume the values 1 and -1, respectively, for the actual and desired joint variables. The number of neurons in the hidden layer plays an important role in the effectiveness of the neural networks as function approximators. From the study presented in [3], confirmed here by the experimental results, seven hidden layer neurons present the best results in terms of the parameter estimation process. The matrix Ξ can be computed as:

$$\Xi = \begin{bmatrix} \xi_1^T & 0 & 0 \\ 0 & \xi_2^T & 0 \\ 0 & 0 & \xi_3^T \end{bmatrix},$$

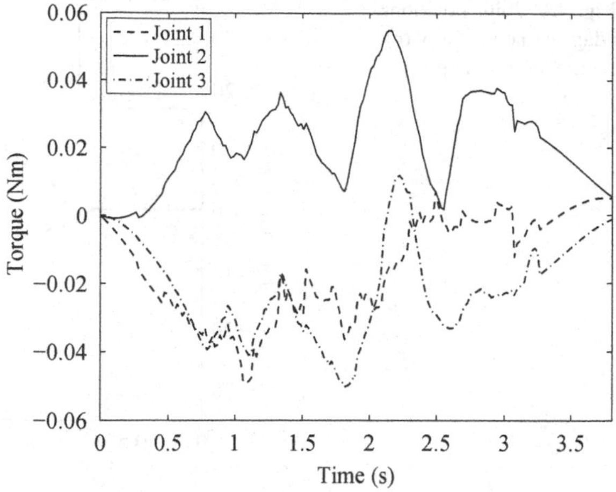

Fig. 4.3 Applied torques, adaptive nonlinear $\mathcal{H}_\infty$ control

where $\xi_1^T = [\xi_{11}, \ldots, \xi_{17}]$, $\xi_2^T = [\xi_{21}, \ldots, \xi_{27}]$ and $\xi_3^T = [\xi_{31}, \ldots, \xi_{37}]$, with ξ_{ki}, $i = 1, \ldots, p_k$, computed according to (4.8). The biases m_i^k assume the values -1.5, -1, -0.5, 0, 0.5, 1 and 1.5, respectively, from the first to seventh neurons, for all neural networks. The network parameters Θ are defined as:

$$\Theta = \begin{bmatrix} \theta_1 \\ \theta_2 \\ \theta_3 \end{bmatrix},$$

with

$$\begin{aligned} \theta_1 &= [\theta_{11}\, \theta_{12}\, \theta_{13}\, \theta_{14}\, \theta_{15}\, \theta_{16}\, \theta_{17}]^T, \\ \theta_2 &= [\theta_{21}\, \theta_{22}\, \theta_{23}\, \theta_{24}\, \theta_{25}\, \theta_{26}\, \theta_{27}]^T, \\ \theta_3 &= [\theta_{31}\, \theta_{32}\, \theta_{33}\, \theta_{34}\, \theta_{35}\, \theta_{36}\, \theta_{37}]^T. \end{aligned}$$

To apply the variable structure controller, we assume that the approximation error is bounded by the state-dependent function $k(x_e)$ defined as:

$$k(x_e) = 2\sqrt{\tilde{x}_1^2 + \tilde{x}_2^2}. \tag{4.12}$$

The results for the adaptive neural network nonlinear $\mathcal{H}_\infty$ control, with $\Theta(0) = [0 \cdots 0]_{21\times 1}^T$, $Z = 10$, and γ, R and Q as defined in Sect. 4.4.2, are shown in Figs. 4.4 and 4.5.

We compare the performance of the two adaptive nonlinear $\mathcal{H}_\infty$ controllers by looking at the values of the performance indexes $\mathcal{L}_2[\tilde{x}]$ and $E[\tau]$. As in Chap. 3, five experiments were performed with each controller and an average of the indexes was computed. The experimental results shown in Figs. 4.2, 4.3, 4.4 and 4.5 correspond to the samples that are closest to the mean values of $\mathcal{L}_2(\tilde{x})$ and $E[\tau]$

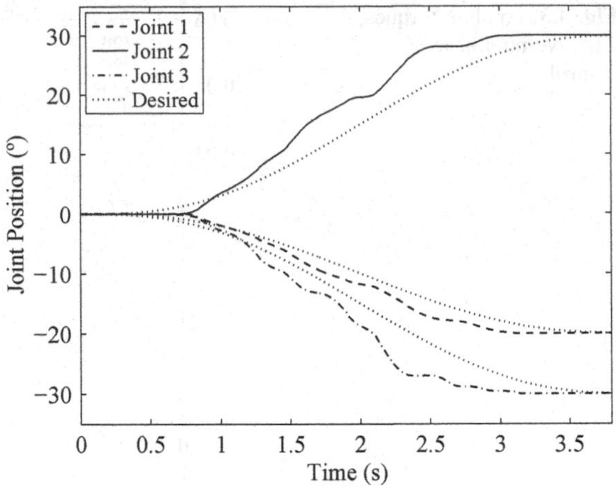

Fig. 4.4 Joint positions, adaptive neural network nonlinear $\mathcal{H}_\infty$ control

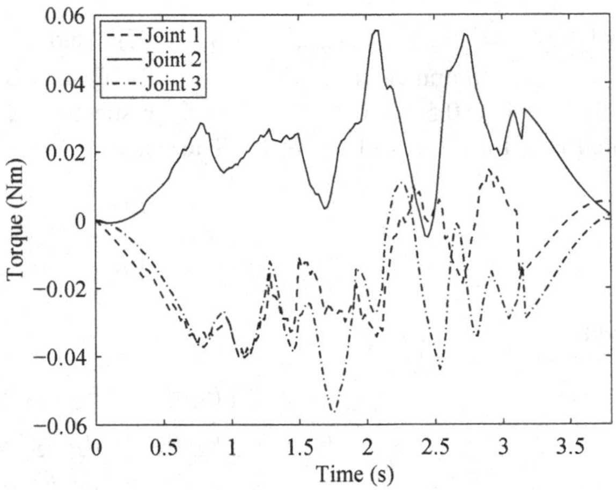

Fig. 4.5 Applied torques, adaptive neural network nonlinear $\mathcal{H}_\infty$ control

Table 4.1 Performance indexes

Controller	$\mathcal{L}_2[\tilde{x}]$	$E[\tau]$ (N m s)
Adaptive nonlinear $\mathcal{H}_\infty$	0.0245	0.2235
Adaptive neural network nonlinear $\mathcal{H}_\infty$	0.0227	0.2257

(Table 4.1). Note that the adaptive neural network nonlinear $\mathcal{H}_\infty$ control presents the best result with respect to the $\mathcal{L}_2$ performance index. However, the total energy spent, indicated by the total torque index, is greater for this controller than for the adaptive $\mathcal{H}_\infty$ control based on linear parameterization. This is natural since the controller needs to devote an extra effort to more efficiently attenuate the effects of the external disturbances on the tracking error. The difference, however, is all but negligible.

References

1. Bergerman M (1996) Dynamics and control of underactuated manipulators. Ph.D. Thesis, Carnegie Mellon University, Pittsburgh, p 129
2. Chang YC (2000) Neural network-based $\mathcal{H}_\infty$ tracking control for robotic systems. IEE Proc Control Theory Appl 147(3):303–311
3. Chang YC, Chen BS (1997) A nonlinear adaptive $\mathcal{H}_\infty$ tracking control design in robotic systems via neural networks. IEEE Trans Control Syst Technol 5(1):13–29
4. Chen BS, Chang YC, Lee TC (1997) Adaptive control in robotic systems with $\mathcal{H}_\infty$ tracking performance. Automatica 33(2):227–234
5. Chen BS, Lee TS, Feng JH (1994) A nonlinear $\mathcal{H}_\infty$ control design in robotic systems under parameter perturbation and external disturbance. Int J Control 59(2):439–461
6. Craig JJ (1985) Adaptive control of mechanical manipulators. Addison-Wesley, Reading
7. Ge SS, Lee TH, Harris CJ (1998) Adaptive neural network control of robotic manipulators. World Scientific, Singapore
8. Lewis FL, Abdallah CT, Dawson DM (2004) Robot manipulator control: theory and practice. Marcel Dekker, Inc., New York
9. Postlethwaite I, Bartoszewicz A (1998) Application of non-linear $\mathcal{H}_\infty$ control to the Tetrabot robot manipulator. Proc Inst Mech Eng: Part I. J Syst Control Eng 212(16):459–465
10. Siqueira AAG, Petronilho A, Terra MH (2003) Adaptive nonlinear $\mathcal{H}_\infty$ techniques applied to a robot manipulator. In: Proceedings of the IEEE conference on control applications, Istanbul, Turkey
11. Slotine JJ, Li W (1987) On the adaptive control of robot manipulators. Int J Rob Res 6(3):49–59

Part II
Fault Tolerant Control of Robot Manipulators

Chapter 5
Underactuated Robot Manipulators

5.1 Introduction

In this book we use the term underactuated manipulators to denote open, serial chain robotic manipulators with more joints than actuators. In general, underactuation can occur as a result of failures, as a consequence of the system's mechanical design, or as an inherent property of the system. Examples of manipulators in each category include: regular industrial manipulator with a failed joint motor; hyper-redundant snake-like robots purposely designed with some unactuated joints; and space manipulators mounted on free-floating satellites. Several references present the particularities of each type of underactuation; a representative sample includes [1–7]. Underactuation usually introduces nonholonomic constraints in the system's dynamic equation. These are non-integrable constraints involving the system state's first- or second-order (or higher) derivatives that allow one to control more degrees of freedom (DOF) than the number of actuators available. (According to [8], the term *holonomous* was coined by Heinrich Hertz (1857–1894)).

Although it may not seem obvious at first, in many cases it is possible to control the position of all joints of an underactuated manipulator thanks to the dynamic coupling between the joints. Here, we consider that the unactuated joints are equipped with on/off brakes. Controling the positions of all joints of an underactuated manipulator is a multi-step process. First, unactuated joints are unlocked and their positions controlled via their dynamic coupling with the actuated ones. As they reach their target positions, they are locked in place. Once all unactuated joints have converged to their desired positions, and are all locked, the actuated joints can be controlled as if the manipulator were fully actuated. In this chapter we present this phased approach in more detail; the number of control phases necessary to control the position of all joints depends on the number of actuated and unactuated joints in the system.

The fact that not all joints are actuated may reduce the system's robustness against disturbances and modeling uncertainties. This represents the main

A. A. G. Siqueira et al., *Robust Control of Robots*,
DOI: 10.1007/978-0-85729-898-0_5, © Springer-Verlag London Limited 2011

motivation for this chapter: to improve the robustness of underactuated manipulators through nonlinear $\mathcal{H}_\infty$ techniques. We start by defining in Sect. 5.2 three different ways of grouping an underactuated manipulator's active and passive joints into appropriate vectors that allow us to cast its dynamic model in a quasi-linear parameter varying (quasi-LPV) form. Based on this quasi-LPV model, we present in Sect. 5.3 an $\mathcal{H}_\infty$ control method via linear matrix inequalities (LMIs) that can be used to design the controller's gains. Then in Sect. 5.4 we present an $\mathcal{H}_\infty$-based controller via game theory that can be used to independently control the actuated and unactuated joints. In Sect. 5.5 we present an alternative adaptive $\mathcal{H}_\infty$ control methodology based on the mathematic model of the underactuated manipulator and on neural networks. Finally, in Sect. 5.6 we present examples of the application of these controllers to the UARM in a configuration with one passive and two active joints.

5.2 Quasi-LPV Representation of Underactuated Manipulators

Consider a manipulator with n joints, of which n_a are actuated and n_p are not, where $n_a + n_p = n$. The unactuated joints are equipped with on/off brakes and are termed *passive* joints; the actuated ones are termed *active* joints. At any given instant, up to n_a joints can be controlled simultaneously [1]. The n_a joints being controlled are grouped in the vector $q_c \in \Re^{n_a}$, the vector of *controlled* joints. All others are grouped in the vector $q_r \in \Re^{n_p}$, the vector of *remaining* joints. There are three possible ways to construct the vector q_c [2]:

1. q_c contains only passive joints. All passive joints not in q_c (if any) are kept locked.
2. q_c contains both passive and active joints. Again, all passive joints not in q_c (if any) are kept locked.
3. q_c contains only active joints. All passive joints are kept locked.

With this in mind, the manipulator's dynamic equation

$$\tau + \delta(q,\dot{q},\ddot{q}) = M_0(q)\ddot{q} + C_0(q,\dot{q})\dot{q} + F_0(\dot{q}) + G_0(q) \tag{5.1}$$

can be partitioned as:

$$\begin{aligned}\begin{bmatrix}\tau_a\\0\end{bmatrix} + \begin{bmatrix}\delta_a(q,\dot{q},\ddot{q})\\\delta_u(q,\dot{q},\ddot{q})\end{bmatrix} &= \begin{bmatrix}M_{ac}(q) & M_{ar}(q)\\M_{uc}(q) & M_{ur}(q)\end{bmatrix}\begin{bmatrix}\ddot{q}_c\\\ddot{q}_r\end{bmatrix}\\&\quad + \begin{bmatrix}C_{ac}(q,\dot{q}) & C_{ar}(q,\dot{q})\\C_{uc}(q,\dot{q}) & C_{ur}(q,\dot{q})\end{bmatrix}\begin{bmatrix}\dot{q}_c\\\dot{q}_r\end{bmatrix} + \begin{bmatrix}F_a(\dot{q})\\F_u(\dot{q})\end{bmatrix} + \begin{bmatrix}G_a(q)\\G_u(q)\end{bmatrix},\end{aligned} \tag{5.2}$$

where the indices a and u represent the active and free (unlocked) passive joints, respectively. For simplicity, the index 0 representing the nominal system is omitted. The matrices with indices *ac*, *ar*, *uc* and *ur* relate torques and accelerations in the active and passive joints to those in the controlled and remaining joints. For example, M_{ac} transforms torque in the active joints to accelerations in the controlled joints, and is at the core of the dynamic coupling-based multi-step control approach described above.

Factoring $\ddot{q}_r$ in the second line of (5.2) and substituting the result back in its first line, we obtain:

$$\tau_a + \overline{\delta}(q,\dot{q},\ddot{q}) = \overline{M}_0(q)\ddot{q}_c + \overline{C}_0(q,\dot{q})\dot{q}_c + \overline{D}_0(q,\dot{q})\dot{q}_r + \overline{F}_0(q,\dot{q}) + \overline{G}_0(q), \tag{5.3}$$

where

$$\begin{aligned}
\overline{M}_0(q) &= M_{ac}(q) - M_{ar}(q)M_{ur}^{-1}(q)M_{uc}(q),\\
\overline{C}_0(q,\dot{q}) &= C_{ac}(q,\dot{q}) - M_{ar}(q)M_{ur}^{-1}(q)C_{uc}(q,\dot{q}),\\
\overline{D}_0(q,\dot{q}) &= C_{ar}(q,\dot{q}) - M_{ar}(q)M_{ur}^{-1}(q)C_{ur}(q,\dot{q}),\\
\overline{F}_0(q,\dot{q}) &= F_a(\dot{q}) - M_{ar}(q)M_{ur}^{-1}(q)F_u(\dot{q}),\\
\overline{G}_0(q) &= G_a(q) - M_{ar}(q)M_{ur}^{-1}(q)G_u(q),\\
\overline{\delta}(q,\dot{q},\ddot{q}) &= \delta_a(q,\dot{q},\ddot{q}) - M_{ar}(q)M_{ur}^{-1}(q)\delta_u(q,\dot{q},\ddot{q}).
\end{aligned}$$

Recall that the objective is to control the positions of the joints in q_c via their dynamic coupling with q_a, irrespective of the nature of the joints in q_c (active or passive). Toward this goal the state tracking error can be defined as:

$$\tilde{x}_c = \begin{bmatrix} q_c - q_c^d \\ \dot{q}_c - \dot{q}_c^d \end{bmatrix} = \begin{bmatrix} \tilde{q}_c \\ \dot{\tilde{q}}_c \end{bmatrix}, \tag{5.4}$$

where q_c^d and $\dot{q}_c^d \in \Re^{n_a}$ are respectively the desired reference position and velocity of the controlled joints. There is no reference position defined for the remaining joints. Hence, a quasi-linear parameter varying (quasi-LPV) representation of the underactuated manipulator can be defined as follows:

$$\begin{aligned}
\dot{\tilde{x}}_c &= \overline{A}(q,\dot{q})\tilde{x}_c + \overline{B}\overline{u} + \overline{B}\overline{w},\\
y &= \overline{C}_1\tilde{x}_c,\\
z &= \overline{C}_2\tilde{x}_c + \overline{u},
\end{aligned} \tag{5.5}$$

where

$$\begin{aligned}
\overline{A}(q,\dot{q}) &= \begin{bmatrix} 0 & I_{n_a} \\ 0 & -\overline{M}_0^{-1}(q)\overline{C}_0(q,\dot{q}) \end{bmatrix},\\
\overline{B} &= \begin{bmatrix} 0 \\ I_{n_a} \end{bmatrix},\\
\overline{w} &= \overline{M}_0^{-1}(q)\overline{\delta}(q,\dot{q},\ddot{q}),\\
\overline{u} &= \overline{M}_0^{-1}(q)(\tau_a - \overline{M}_0(q)\ddot{q}_c^d - \overline{C}_0(q,\dot{q})\dot{q}_c^d - \overline{D}_0(q,\dot{q})\dot{q}_r - \overline{F}_0(\dot{q}) - \overline{G}_0(q)).
\end{aligned}$$

From (5.5), the applied torque is given by:

$$\tau_a = \overline{M}_0(q)\ddot{q}_c^d + \overline{C}_0(q,\dot{q})\dot{q}_c^d + \overline{D}_0(q,\dot{q})\dot{q}_r + \overline{F}_0(\dot{q}) + \overline{G}_0(q) + \overline{M}_0(q)\overline{u}. \quad (5.6)$$

As in Chap. 3, although the matrix $\overline{M}_0(q)$ explicitly depends on the joint positions, we can consider it as function of the position error $\overline{M}_0(q) = \overline{M}_0(q_c, q_r) = \overline{M}_0(\tilde{q}_c + q_c^d(t), q_r) = \overline{M}_0(\tilde{x}_c, q_r, t)$. The same can be considered for $\overline{C}_0(q,\dot{q})$, except that $\overline{C}_0(q,\dot{q})$ is also a function of $\dot{q}_r$.

5.3 $\mathcal{H}_\infty$ Control via Linear Matrix Inequalities

The role of the $\mathcal{H}_\infty$ controller is to guarantee that the controlled joints reach their desired positions while holding the following inequality true:

$$\int_0^{T_f} \|z(t)\|^2 dt \leq \gamma^2 \int_0^{T_f} \|\overline{w}(t)\|^2 dt, \quad (5.7)$$

for all $T_f \geq 0$ and all $\overline{w} \in \mathcal{L}_2(0, T_f)$ with the system starting from $\tilde{x}_c(0) = 0$. The parameter γ in this inequality assumes a role equivalent to that defined in the standard $\mathcal{H}_\infty$ criterion for linear systems: it establishes a level of attenuation of the input disturbances on the output of the system. In this section we assume that all joints are equipped with the appropriate sensors to measure joint position and velocity. This assumption will be relaxed in future chapters when we study fault-tolerance control for underactuated manipulators. The state feedback control problem considered here aims to find a continuous function $F(\rho(\tilde{x}_c))$ such that the closed loop system has an $\mathcal{L}_2$ gain less than or equal to γ under a state feedback law $u = F(\rho(\tilde{x}_c))\tilde{x}_c$, where $\rho(\tilde{x}_c)$ belongs to the set defined in Eq. 3.30.

According to [9] and [10], if there exists a continuously differentiable matrix function $X(\rho(\tilde{x}_c)) > 0$ that satisfies

$$\begin{bmatrix} \overline{E}(\rho(\tilde{x}_c)) & X(\rho(\tilde{x}_c))\overline{C}_1^T & \overline{B} \\ \overline{C}_1 X(\rho(\tilde{x}_c)) & -I & 0 \\ \overline{B}^T & 0 & -\gamma^2 I \end{bmatrix} < 0, \quad (5.8)$$

where

$$\overline{E}(\rho(\tilde{x}_c)) = -\sum_{i=1}^{m} \underline{\overline{v}}_i(\rho)\frac{\partial X(\rho(\tilde{x}_c))}{\partial \rho_i} - \overline{B}\overline{B}^T + \widehat{\overline{A}}(\rho(\tilde{x}_c))X(\rho(\tilde{x}_c)) + X(\rho(\tilde{x}_c))\widehat{\overline{A}}(\rho(\tilde{x}_c))^T,$$

and $\widehat{\overline{A}}(\rho(x_c)) = \overline{A}(\rho(\tilde{x}_c)) - \overline{BC}_2$, then the closed loop system has $\mathcal{L}_2$ gain $\leq \gamma$ under the state feedback control law:

$$u(t) = -(\overline{B}X^{-1}(\rho(\tilde{x}_c)) + \overline{C}_2)\tilde{x}_c(t).$$

For the underactuated control case, we define different underlying parameters $\rho(\tilde{x}_c)$ according to the nature of the joints being controlled. In each control phase, the combination of the positions and velocities of the manipulator is given by the summation $\sum_{i=1}^{m} \underline{\overline{v}}_i(\rho(\tilde{x}_c))$, which represents that every combination of $\overline{v}_i(\rho(\tilde{x}_c))$ and $\underline{v}_i(\rho(\tilde{x}_c))$ should be included in the inequality (5.8) (recal that $\underline{v}_i(\rho) \leq \dot{\rho}_i \leq \overline{v}_i(\rho), i = 1, \ldots, m$). The solutions of each set of LMIs, for each control phase, follow the same reasoning described in Sect. 3.4. First, select a set of basis functions $\{f_i(\rho(\tilde{x}_c))\}_{i=1}^{M}$ for $X(\rho(\tilde{x}_c))$ and rewrite it as:

$$X(\rho(\tilde{x}_c)) = \sum_{i=1}^{M} f_i(\rho(\tilde{x}_c))X_i, \tag{5.9}$$

where $X_i \in S^{n\times n}$ is the coefficient matrix for $f_i(\rho(\tilde{x}_c))$. Inserting $X(\rho(\tilde{x}_c))$ in (5.8), the constraints turn into an LMI in terms of the matrix variables $\{X_i\}_{i=1}^{M}$, when the parameter $\rho(\tilde{x}_c)$ is fixed, and we can define the following optimization problem

$$\min_{\{X_i\}_{i=1}^{M}} \gamma^2$$

subject to

$$\begin{bmatrix} E^*(\rho(\tilde{x}_c)) & \sum_{j=1}^{M} f_j(\rho(\tilde{x}_c))X_j C_1^T(\rho(\tilde{x}_c)) & \overline{B}(\rho(\tilde{x}_c)) \\ C_1(\rho(\tilde{x}_c))\sum_{j=1}^{M} f_j(\rho(\tilde{x}_c))X_j & -I & 0 \\ \overline{B}^T(\rho(\tilde{x}_c)) & 0 & -\gamma^2 I \end{bmatrix} < 0,$$

$$\sum_{j=1}^{M} f_j(\rho(\tilde{x}_c))X_j > 0, \tag{5.10}$$

where

$$E^*(\rho(\tilde{x}_c)) = -\sum_{i=1}^{m} \underline{\overline{v}}_i(\rho(\tilde{x}_c))\frac{\partial X(\rho(\tilde{x}_c))}{\partial \rho_i} + \sum_{j=1}^{M} f_j(\rho(\tilde{x}_c))(\widehat{A}(\rho(\tilde{x}_c))X_j + X_j\widehat{A}^T(\rho(\tilde{x}_c))) - \overline{B}(\rho(\tilde{x}_c))\overline{B}^T(\rho(\tilde{x}_c)). \tag{5.11}$$

The procedure to solve this optimization problem is identical to the one described in Sect. 3.4, and should be applied to each control phase according to the nature of the joints in the vector q_c. We present a detailed example later in this chapter.

5.4 $\mathcal{H}_\infty$ Control via Game Theory

Fully-actuated manipulator control via game theory requires that $C_0(q,\dot{q}) - \frac{1}{2}\dot{M}_0(q,\dot{q})$ be a skew-symmetric matrix. The partition used in Eq. 5.2, however, does not guarantee that this property is satisfied in the underactuated case. Therefore, we introduce a new partition of Eq. 5.1 that maintains that matrix's skew-symmetry:

$$\begin{bmatrix} \tau_c \\ \tau_r \end{bmatrix} + \begin{bmatrix} \delta_c(q,\dot{q},\ddot{q}) \\ \delta_r(q,\dot{q},\ddot{q}) \end{bmatrix} = \begin{bmatrix} M_{cc}(q) & M_{cr}(q) \\ M_{rc}(q) & M_{rr}(q) \end{bmatrix} \begin{bmatrix} \ddot{q}_c \\ \ddot{q}_r \end{bmatrix} + \begin{bmatrix} C_{cc}(q,\dot{q}) & C_{cr}(q,\dot{q}) \\ C_{rc}(q,\dot{q}) & C_{rr}(q,\dot{q}) \end{bmatrix} \begin{bmatrix} \dot{q}_c \\ \dot{q}_r \end{bmatrix} + \begin{bmatrix} F_c(\dot{q}) \\ F_r(\dot{q}) \end{bmatrix} + \begin{bmatrix} G_c(q) \\ G_r(q) \end{bmatrix}, \tag{5.12}$$

where τ_c are the controlled joint torques and τ_r are the remaining joint torques. As we have shown in Chap. 3, we solve the $\mathcal{H}_\infty$ underactuated manipulator control problem via game theory through the following state transformation:

$$\tilde{z} = \begin{bmatrix} \tilde{z}_1 \\ \tilde{z}_2 \end{bmatrix} = T_0 \tilde{x}_c = \begin{bmatrix} T_1 \\ T_2 \end{bmatrix} \tilde{x}_c = \begin{bmatrix} I & 0 \\ T_{11} & T_{12} \end{bmatrix} \begin{bmatrix} \tilde{q}_c \\ \dot{\tilde{q}}_c \end{bmatrix}, \tag{5.13}$$

where $T_{11}, T_{12} \in \Re^{n \times n}$ are constant matrices to be determined. The control input can then be selected as:

$$\overline{u} = M_{cc}(q) T_2 \dot{\tilde{x}}_c + C_{cc}(q,\dot{q}) T_2 \tilde{x}_c. \tag{5.14}$$

Considering the second line of Eq. 5.12 and the state tracking error (5.4), the state transformation (5.13) can be used to generate the following dynamic equation of the underactuated manipulator:

$$\dot{\tilde{x}}_c = \overline{A}_T(\tilde{x}_c,t)\tilde{x}_c + \overline{B}_T(\tilde{x}_c,t)\overline{u} + \overline{B}_T(\tilde{x}_c,t)\overline{w}, \tag{5.15}$$

where

$$\overline{A}_T(\tilde{x}_c,t) = T_0^{-1} \begin{bmatrix} -T_{12}^{-1}T_{11} & T_{12}^{-1} \\ 0 & -M_{cc}^{-1}(q)C_{cc}(q,\dot{q}) \end{bmatrix} T_0,$$

$$\overline{B}_T(\tilde{x}_c,t) = T_0^{-1} \begin{bmatrix} 0 \\ M_{cc}^{-1}(q) \end{bmatrix},$$

$$\overline{w} = M_{cc}(q) T_{12} M_{cc}^{-1}(q) \delta_c(q,\dot{q},\ddot{q}).$$

Note that we are considering $M_{cc}(q)$ in terms of controlled positions, and as a consequence the matrix $A_T(\tilde{x}_c, t)$ is also given in terms of controlled positions. From Eq. 5.14, the control acceleration is given by:

$$\ddot{q}_c = \ddot{q}_c^d - T_{12}^{-1}T_{11}\dot{\tilde{x}}_c - T_{12}^{-1}M_{cc}^{-1}(q)\big(C_{cc}(q,\dot{q})B^T T_0\tilde{x}_c - \bar{u}\big). \tag{5.16}$$

Equation 5.16 gives the necessary accelerations for the controlled joints to follow the desired reference trajectory, from which the torques in the active joints can be calculated. First, rewrite Eq. 5.2 as:

$$\begin{bmatrix}\tau_a \\ 0\end{bmatrix} + \begin{bmatrix}\delta_a(q,\dot{q},\ddot{q}) \\ \delta_u(q,\dot{q},\ddot{q})\end{bmatrix} = \begin{bmatrix}M_{ac}(q) & M_{ar}(q) \\ M_{uc}(q) & M_{ur}(q)\end{bmatrix}\begin{bmatrix}\ddot{q}_c \\ \ddot{q}_r\end{bmatrix} + \begin{bmatrix}b_a(q,\dot{q}) \\ b_u(q,\dot{q})\end{bmatrix}, \tag{5.17}$$

where $b(q,\dot{q}) = C(q,\dot{q}) + F(\dot{q}) + G(q)$; then factor out $\ddot{q}_r$ in the second line and substitute it in the first one to obtain

$$\begin{aligned}\tau_a = {} & \big(M_{ac}(q) - M_{ar}(q)M_{ur}^{-1}(q)M_{uc}(q)\big)\ddot{q}_c + b_a(q,\dot{q}) \\ & - \delta_a(q,\dot{q},\ddot{q}) - M_{ar}(q)M_{ur}^{-1}(q)(b_u(q,\dot{q}) - \delta_u(q,\dot{q},\ddot{q})).\end{aligned} \tag{5.18}$$

The optimal control law for the underactuated case is based on the same theoretical arguments developed for the fully-actuated case (see Chap. 3 for more details):

$$\bar{u}^* = -R^{-1}\bar{B}^T T_0\tilde{x}_c, \tag{5.19}$$

where the matrix $\bar{B}$ is given in Eq. 5.5. The matrix $P_c(\tilde{x}_c, t)$ is given by

$$P_c(\tilde{x}_c, t) = T_0^T\begin{bmatrix}K_c & 0 \\ 0 & M_{cc}(\tilde{x}_c, t)\end{bmatrix}T_0,$$

where K_c is a positive definite symmetric constant matrix which solves the analogous $\mathcal{H}_\infty$ control problem for the underactuated case. The guidelines to select T_0 and K_c for each control phase follow the approach described in Sect. 3.3. Note that the skew-symmetry of the matrix $C_{cc}(q,\dot{q}) - \frac{1}{2}\dot{M}_{cc}(q,\dot{q})$ is guaranteed, as is necessary in this formulation.

5.5 Adaptive $\mathcal{H}_\infty$ Control

In this section, the adaptive $\mathcal{H}_\infty$ control methodologies described in Chap. 4 are applied to the underactuated manipulators. Here too it is important that matrix $C(q,\dot{q}) - \frac{1}{2}\dot{M}(q,\dot{q})$ be skew-symmetric, which holds when partition (5.12) is used. When dealing with adaptive controllers, one must find a state space equation relating the control input to the torque applied to the active joints; this is the subject of the next sections.

5.5.1 Model-Based Control

To apply the adaptive control methods to the passive joints of an underactuated manipulator, we must find a new representation of Eq. 5.1. For convenience, the Coriolis and centripetal torques are represented as: $D(q,\dot{q})\dot{q}$, with $D(q,\dot{q}) \in \Re^{nxn}$. With q_c composed only by passive joints, Eq. 5.1 can be partitioned as:

$$\begin{bmatrix} 0 \\ \tau_r \end{bmatrix} + \begin{bmatrix} \delta_c \\ \delta_r \end{bmatrix} = \begin{bmatrix} M_{cc}(q) & M_{cr}(q) \\ M_{rc}(q) & M_{rr}(q) \end{bmatrix} \begin{bmatrix} \ddot{q}_c \\ \ddot{q}_r \end{bmatrix} + \begin{bmatrix} D_{cc}(q,\dot{q}) & D_{cr}(q,\dot{q}) \\ D_{rc}(q,\dot{q}) & D_{rr}(q,\dot{q}) \end{bmatrix} \begin{bmatrix} \dot{q}_c \\ \dot{q}_r \end{bmatrix} + \begin{bmatrix} F_c(\dot{q}) \\ F_r(\dot{q}) \end{bmatrix} + \begin{bmatrix} G_c(q) \\ G_r(q) \end{bmatrix}, \tag{5.20}$$

where $\tau_c = 0$ because only the passive joints are being controlled, and the dependence of δ on $(q,\dot{q},\ddot{q})$ is omitted for brevity. Factoring out $\ddot{q}_r$ in the first line of Eq. 5.20 and substituting it in the second one we obtain:

$$\tau_a + \overline{\delta} = \overline{M}(q)\ddot{q}_c + \overline{D}(q,\dot{q})\dot{q}_c + \overline{E}(q,\dot{q})\dot{q}_r + \overline{F}(\dot{q}) + \overline{G}(q), \tag{5.21}$$

with $\tau_a = \tau_r$ and

$$\begin{aligned}
\overline{M}(q) &= M_{rc}(q) - M_{rr}(q)M_{cr}^{-1}(q)M_{cc}(q),\\
\overline{D}(q,\dot{q}) &= D_{rc}(q,\dot{q}) - M_{rr}(q)M_{cr}^{-1}(q)D_{cc}(q,\dot{q}),\\
\overline{E}(q,\dot{q}) &= D_{rr}(q,\dot{q}) - M_{rr}(q)M_{cr}^{-1}(q)D_{cr}(q,\dot{q}),\\
\overline{F}(\dot{q}) &= F_r(\dot{q}) - M_{rr}(q)M_{cr}^{-1}(q)F_c(\dot{q}),\\
\overline{G}(q) &= G_r(q) - M_{rr}(q)M_{cr}^{-1}(q)G_c(q),\\
\overline{\delta} &= \delta_r(q,\dot{q},\ddot{q}) - M_{rr}(q)M_{cr}^{-1}(q)\delta_c(q,\dot{q},\ddot{q}).
\end{aligned}$$

Considering (5.21), the state tracking error (5.4), and the state transformation (5.13), a new dynamic equation of the underactuated manipulator can be written as:

$$\dot{\tilde{x}}_c = \overline{A}_T(\tilde{x}_c,t)\tilde{x}_c + \overline{B}_T(\tilde{x}_c,t)T_{12}(-\overline{F}_0(x_e) + \tau_a) + \overline{B}_T(\tilde{x}_c,t)\overline{d}, \tag{5.22}$$

with

$$\begin{aligned}
\overline{A}_T(\tilde{x}_c,t) &= T_0^{-1}\begin{bmatrix} -T_{12}^{-1}T_{11} & T_{12}^{-1} \\ 0 & -\overline{M}_0(q)^{-1}\overline{D}_0(q,\dot{q}) \end{bmatrix} T_0,\\
\overline{B}_T(\tilde{x}_c,t) &= T_0^{-1}\begin{bmatrix} 0 \\ \overline{M}_0^{-1}(q) \end{bmatrix},\\
\overline{d} &= \overline{M}_0(q)T_{12}\overline{M}_0^{-1}(q)(\overline{\delta} + \Delta\overline{F}(x_e)),
\end{aligned}$$

$$
\begin{aligned}
\overline{F}_0(x_e) &= \overline{M}_0(q)(\ddot{q}_c^d - T_{12}^{-1}T_{11}\dot{\tilde{q}}_c) + \overline{D}_0(q,\dot{q})(\dot{q}_c^d - T_{12}^{-1}T_{11}\tilde{q}_c) \\
&\quad + \overline{E}_0(q,\dot{q})\dot{q}_r + \overline{F}_0(\dot{q}) + \overline{G}_0(q), \\
\Delta\overline{F}(x_e) &= \Delta\overline{M}(q)(\ddot{q}_c^d - T_{12}^{-1}T_{11}\dot{\tilde{q}}_c) + \Delta\overline{D}(q,\dot{q})(\dot{q}_c^d - T_{12}^{-1}T_{11}\tilde{q}_c) \\
&\quad + \Delta\overline{E}(q,\dot{q})\dot{q}_r + \Delta\overline{F}(\dot{q}) + \Delta\overline{G}(q),
\end{aligned}
$$

where $\overline{M}_0(q)$, $\overline{D}_0(q,\dot{q})$, $\overline{E}_0(q,\dot{q})$, $\overline{F}_0(\dot{q})$, and $\overline{G}_0(q)$ are the nominal variables of Eq. 5.20. Note that we added in this equation parametric uncertainty variables $\Delta\overline{M}(q)$, $\Delta\overline{D}(q,\dot{q})$, $\Delta\overline{E}(q,\dot{q})$, $\Delta\overline{F}(\dot{q})$, and $\Delta\overline{G}(q)$ related with $\overline{M}(q)$, $\overline{D}(q,\dot{q})$, $\overline{E}(q,\dot{q})$, $\overline{F}(\dot{q})$, and $\overline{G}(q)$, respectively.

When only passive joints are controlled, the inertia matrix $\overline{M}_0(q)$ will always be negative definite. In order to solve the nonlinear $\mathcal{H}_\infty$ control problem for underactuated manipulators, following the reasoning presented in Sect. 5.4, the solution of the related Riccati equation, $P_c(\tilde{x}_c,t)$, must be a positive definite symmetric matrix for all $\tilde{x}_c$ and t. Here, $P_c(\tilde{x}_c,t)$ is selected as:

$$
P_c(\tilde{x}_c,t) = T_0^T \begin{bmatrix} K_c & 0 \\ 0 & -\overline{M}_0(\tilde{x}_c,t) \end{bmatrix} T_0,
$$

where K_c is a positive definite symmetric constant matrix. Additionally, to simplify the Riccati equation the matrix $\overline{D}(q,\dot{q})$ must be such that $\overline{N}(q,\dot{q}) = \overline{D}(q,\dot{q}) - \frac{1}{2}\dot{\overline{M}}(q,\dot{q})$ is skew-symmetric. This can be satisfied with a suitable definition of the matrices $D_{cc}(q,\dot{q})$ and $D_{rc}(q,\dot{q})$ resulting from the partition of $D(q,\dot{q})$ in Eq. 5.21. The other entries of $D(q,\dot{q})$, $D_{rr}(q,\dot{q})$ and $D_{cr}(q,\dot{q})$, are determined such that $D(q,\dot{q})\dot{q} = V(q,\dot{q})$. The applied torque in the active joints for the nonlinear $\mathcal{H}_\infty$ control is given by:

$$
\tau_r = \overline{F}_0(x_e) + T_{12}^{-1}R^{-1}B^T T_0 \tilde{x}_c.
$$

For the adaptive nonlinear $\mathcal{H}_\infty$ control, $\Delta\overline{F}(x_e)$ can be written as:

$$
\Delta\overline{F}(x_e) = \overline{Y}(q,\dot{q},\dot{q}_c^d - T_{12}T_{11}\tilde{q}_c, \ddot{q}_c^d - T_{12}^{-1}T_{11}\dot{\tilde{q}}_c)\overline{\theta}. \tag{5.23}
$$

Finally, the adaptive control law is given by:

$$
\begin{aligned}
\dot{\hat{\overline{\theta}}} &= \overline{S}^{-1}\overline{Y}^T T_{12} B^T T_0 \tilde{x}_c, \\
\tau_r &= \overline{F}_0(x_e) + \overline{Y}\hat{\overline{\theta}} + T_{12}^{-1}R^{-1}B^T T_0 \tilde{x}_c.
\end{aligned}
$$

5.5.2 *Neural Network-Based Control*

For the adaptive neural network-based nonlinear $\mathcal{H}_\infty$ control described in Chap. 4, we consider, in the underactuated case, $\Delta\overline{F}(x_e)$ according to (5.8) written as:

$$\Delta\overline{F}(x_e, \overline{\Theta}) = \overline{\Xi\Theta},$$

with $\overline{\Xi}$ and $\overline{\Theta}$ given in Eq. 4.9 of Sect. 4.3.

The nonlinear $\mathcal{H}_\infty$ adaptive control law and the applied torque in the active joints are given by:

$$\dot{\overline{\Theta}} = \overline{Z}^{-T}\overline{\Xi}^T T_{12} B^T T_0 \tilde{x}_c,$$
$$\overline{\tau}_r = \overline{f}_0(x_e, \overline{\Theta}) + \overline{\Xi\Theta} + T_{12}^{-1} R^{-1} B^T T_0 \tilde{x}_c + T_{12}^{-1} \overline{k}(x_e) sgn(B^T T_0 \tilde{x}_c),$$

where $\overline{k}(x_e)$ is a function such that $|(\delta\overline{F}(x_e))_i| \leq \overline{k}(x_e)$, for all $1 \leq i \leq n$, and $\delta\overline{F}(x_e)$ is the approximation error generated by the neural network.

5.6 Examples

To validate the controllers proposed in this chapter we present their application to the three-link planar underactuated manipulator UARM. We consider that one joint is passive and two are active. Accordingly, the multi-step control approach consists of the following phases.

1. Control the position of the passive joint and one active joint until the passive joint reaches its set-point and is locked in place.
2. Control the position of the active joints as if the manipulator were fully-actuated.

5.6.1 *Design Procedure*

The underactuated configuration selected in the tests presented here consists of joints 1 and 3 active and joint 2 passive, or APA for short. In this case $n_a = 2$, and therefore at most two joint positions can be controlled during each control phase. In phase 1, named APA_u, the vector of controlled joints is selected as $q_c = [q_2\ q_3]^T$; in other words, in phase 1 we control the position of the passive joint and one of the active joints. In phase 2, named APA_l, $q_c = [q_1\ q_3]^T$ with joint 2 locked in place as it has already reached its set-point.

From a programming perspective it is important that the matrices in the underactuated manipulator's dynamic equation be appropriately partitioned according to each control phase. The following MATLAB®code presents an example extracted directly from the Underactuated Manipulator Control Environment:

File: uarm_config.m

```
...

elseif (config == '3APA'),
  % 3 links, 2 actuators at joints 1 and 3
  na = 2;
  active = [1 3];
  passive = [2];
  remaining = [1];
  controlled = [2 3];
  active_controlled = [3];

elseif (config == '3PAA'),

...
```

File: uarm_loop.m

```
...
% Partition inertia matrix
Mestac = Mest(active,controlled);
Mestur = Mest(passive,remaining);
Mestar = Mest(active,remaining);
Mestuc = Mest(passive,controlled);

Mestaa = Mest(active,active);
Mestcc = Mest(controlled,controlled);
Cestaa = Cest(active,active);
Cestcc = Cest(controlled,controlled);

...
```

File: uarm_cont_GTH.m

```
...

% Control law
u(controlled,i) = -inv(R)*Be'*T0*[errorq(controlled); ...
      errorqd(controlled)];
qcdd_c(:,i) = qdd_d(controlled,i) - inv(T12)*T11* ...
      errorqd(controlled) - inv(T12)*inv(Mestcc)* ...
      (Cestcc*Be'*T0*[errorq(controlled); ...
      errorqd(controlled)] - u(controlled,i));
```

```
% Active joints control passive joints
tau(active,i) = (Mestac - Mestar*inv(Mestur)*Mestuc)* ...
        qcdd_c(:,i) - Mestar*inv(Mestur)*best(passive) + ...
        best(active);

% Passive joints cannot apply torque
tau(passive,i) = 0;

...
```

The proposed controllers can also be designed via the **Controller Design box** of the UMCE, presented in Chap. 3. Figure 5.1 shows the necessary control parameters for the adaptive $\mathcal{H}_\infty$ controller. The controllers for the two control phases of the APA configuration are designed simultaneously, using the guidelines shown in Chaps. 3 and 4.

5.6.2 LMI-Based Control

We implemented the LMI-based position control method described in Sect. 5.3 on the UARM in the APA configuration. In control phase 1 $\rho(\tilde{x}_c)$ is selected as the state representing the position errors of joints 2 and 3, i.e.,

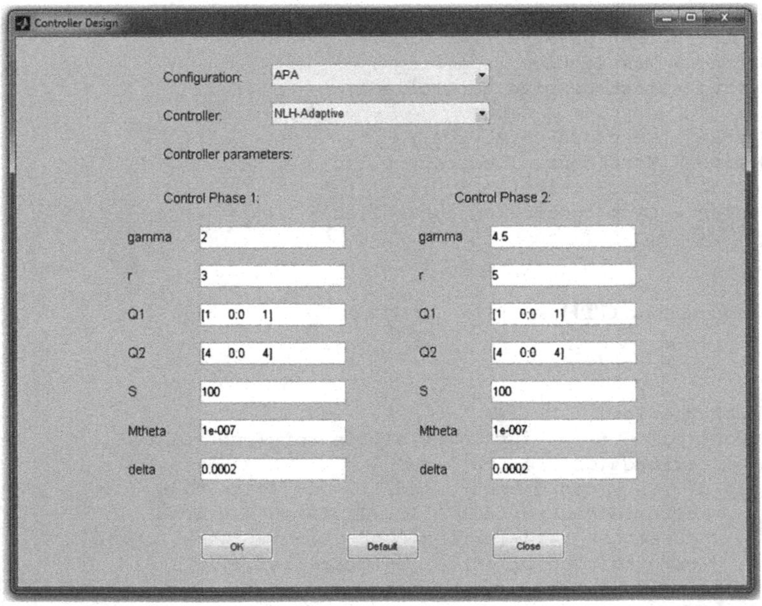

Fig. 5.1 Controller design box for the adaptive $\mathcal{H}_\infty$ controller

$$\rho(\tilde{x}_c) = [\tilde{q}_2 \quad \tilde{q}_3]^T.$$

The system outputs, z_1 and z_2, are the position and velocity errors of the controlled joints and the control variable, u, respectively. Hence, the system can be described by:

$$\begin{aligned}
A(\rho(x)) &= \overline{A}(\rho(\tilde{x}_c)),\\
B_1(\rho(x)) &= \overline{B},\\
B_2(\rho(x)) &= \overline{B},\\
C_1(\rho(x)) &= I_4,\\
C_2(\rho(x)) &= 0,
\end{aligned}$$

where the matrices $\overline{A}(\rho(\tilde{x}_c))$ and $\overline{B}$ are defined in (5.5). The parameter space is divided in $L = 5$ grid points. The best attenuation level is $\gamma = 1.35$. In control phase 2 the state vector is given by:

$$\rho(\tilde{x}_c) = \left[\tilde{q}_3 \quad \dot{\tilde{q}}_3\right]^T,$$

where $\tilde{q}_3$ and $\dot{\tilde{q}}_3$ are the position and velocity errors of joint 3. Again, the position and velocity errors of the controlled joints and the control variable, u, are the outputs of the system. We define $\rho(\tilde{x}_c) \in [-30, 30]^\circ \times [-50, 50]^\circ/\text{s}$. The parameter variation rate is bounded by $|\dot{\rho}| \le [50^\circ/\text{s} \ 30^\circ/\text{s}^2]$. The basis functions selected for both control phases are:

$$\begin{aligned}
f_1(\rho(\tilde{x}_c)) &= 1,\\
f_2(\rho(\tilde{x}_c)) &= \cos(\tilde{q}_3),\\
f_3(\rho(\tilde{x}_c)) &= \cos(\dot{\tilde{q}}_3).
\end{aligned}$$

The parameter space is again divided in $L = 5$ grid points. The best level of attenuation in this phase is $\gamma = 1.80$. The solutions to the LMI problem for the first and second control phases are given by:

$$X_1 = \begin{bmatrix} 0.0893 & -0.0369 & -0.0660 & 0.0725\\ -0.0369 & 0.1359 & 0.0288 & -0.1667\\ -0.0660 & 0.0288 & 0.1693 & -0.0160\\ 0.0725 & -0.1667 & -0.0160 & 0.4121 \end{bmatrix},$$

$$X_2 = \begin{bmatrix} 0.0447 & -0.0112 & -0.0036 & 0.0320\\ -0.0112 & 0.0110 & -0.0085 & -0.0401\\ -0.0036 & -0.0085 & 0.0035 & 0.0121\\ 0.0320 & -0.0401 & 0.0121 & 0.0412 \end{bmatrix},$$

$$X_3 = \begin{bmatrix} -0.0094 & 0.0214 & -0.0152 & -0.0174 \\ 0.0214 & 0.0128 & 0.0158 & 0.0438 \\ -0.0152 & 0.0158 & -0.0087 & -0.0402 \\ -0.0174 & 0.0438 & -0.0402 & -0.0588 \end{bmatrix},$$

and

$$X_1 = \begin{bmatrix} 0.2306 & 0.0115 & -0.1994 & -0.0008 \\ 0.0115 & 0.2177 & -0.0194 & -0.1812 \\ -0.1994 & -0.0194 & 0.3850 & 0.0059 \\ -0.0008 & -0.1812 & 0.0059 & 0.3452 \end{bmatrix},$$

$$X_2 = \begin{bmatrix} -0.0121 & -0.0393 & 0.0151 & 0.0386 \\ -0.0393 & 0.0023 & 0.0386 & -0.0025 \\ 0.0151 & 0.0386 & -0.0206 & 0.0084 \\ 0.0386 & -0.0025 & 0.0084 & 0.0050 \end{bmatrix},$$

$$X_3 = \begin{bmatrix} -0.0174 & 0.0241 & 0.0158 & 0.0051 \\ 0.0241 & 0.0144 & -0.0431 & -0.0055 \\ 0.0158 & -0.0431 & -0.0176 & 0.0183 \\ 0.0051 & -0.0055 & 0.0183 & 0.0037 \end{bmatrix},$$

respectively. The control law is implemented as:

$$u(t) = -\left(\overline{B}\begin{bmatrix} 0 \\ I_{n_a} \end{bmatrix}(X_1 + X_2\cos(\tilde{q}_3) + X_3\cos(\dot{\tilde{q}}_3))^{-1}\right)\tilde{x}_c(t).$$

In the experiment, the initial position and the desired final position adopted were, respectively, $q(0) = [0° \ 0° \ 0°]^T$ and $q(T^1, T^2) = [20° \ 20° \ 20°]^T$, where $T^1 = [1.0 \ 1.0]$ s and $T^2 = [5.0 \ 5.0]$ s are the trajectory duration time for phases 1 and 2, respectively. An external disturbance τ_d of the following form is introduced in joints 1 and 3 starting at 0.3 s:

$$\tau_d = \begin{bmatrix} 0.5e^{-4t}\sin(4\pi t) \\ -0.05e^{-6t}\sin(4\pi t) \end{bmatrix}.$$

The peak of the disturbance signal is approximately equal to 30% of the torque value at $t = 0.3$ s. Figures 5.2 and 5.3 present the resulting joint positions and applied torques.

5.6.3 Game Theory-Based Control

Similar experiments were run with the UARM in configuration APA with the $\mathcal{H}_\infty$ controller via game theory presented in Sect. 5.4. In control phase 1 the best

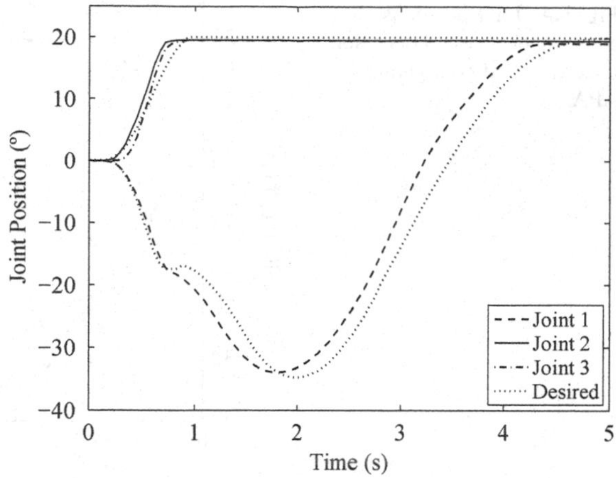

Fig. 5.2 Joint positions, LMI-based controller, underactuated configuration APA

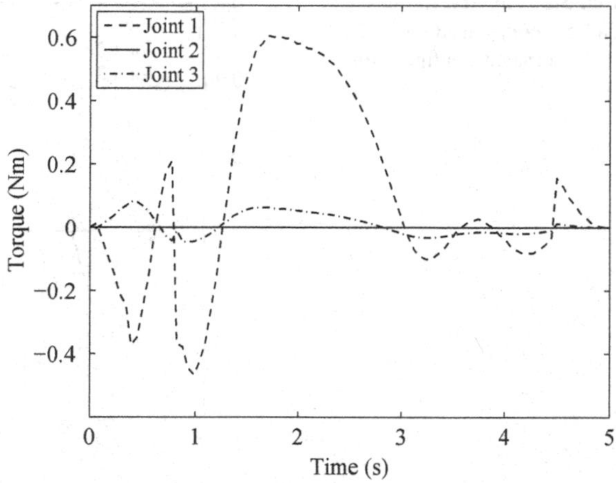

Fig. 5.3 Applied torques, LMI-based controller, underactuated configuration APA

attenuation level is $\gamma = 1.9$ for weighting matrices $Q_1 = 4I_2, Q_2 = I_2, Q_{12} = 0$, and $R = 3.5I_2$. In control phase 2 the best attenuation level is $\gamma = 1.9$ for the same weighting matrices. The experimental results are shown in Figs. 5.4 and 5.5, respectively. Note that in control phase 1, $M_{cc}(q)$ is a function of the positions of joints 2 and 3, as in Eq. 5.15, i.e.,

$$M_{cc}(q) = [M_{22}(q)\ M_{23}(q);\ M_{32}(q)\ M_{33}(q)].$$

In control phase 2, joint 2 is locked and $M_{cc}(q) = [M_{11}(q)\ M_{13}(q);\ M_{31}(q)M_{33}(q)]$ is a function only of q_3, which is a controlled joint in this phase. Hence, matrix $M_{cc}(q)$ is a function only of the controlled joints during both control phases of this configuration (APA).

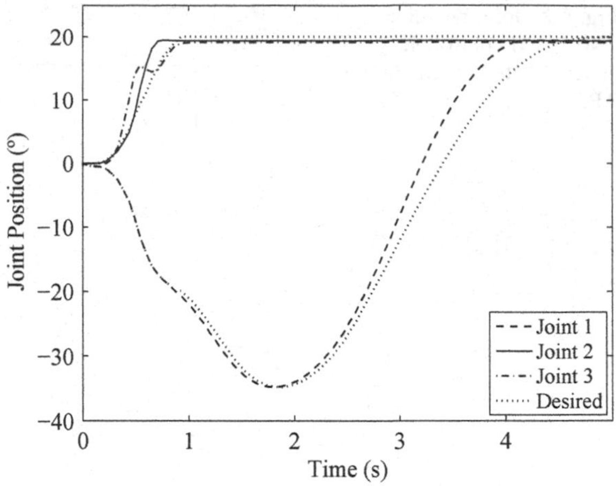

Fig. 5.4 Joint positions, game theory-based controller, underactuated configuration APA

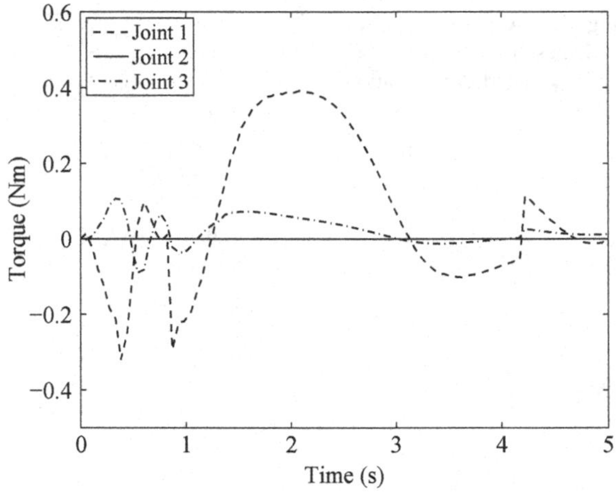

Fig. 5.5 Applied torques, game theory-based controller, underactuated configuration APA

Comparing Figs. 5.2, 5.3 and 5.4,5.5 one can see that the LMI-based controller is more robust to the external disturbance than the game theory-based one. Note, for example, that for the latter the positions of joints 2 and 3 oscillate more in the presence of the disturbance. This is partly due to the fact that the game theory-based control law has a static gain, which makes it less effective in terms of disturbance rejection—similarly to feedback linearization control laws. The greater robustness of the LMI-based controller can also be verified by comparing the values of γ obtained with both methods (Table 5.1). It is important to note, however, that a disadvantage of the LMI-based controller is the more complex design procedure, especially the selection of the basis functions.

The experiments described above were repeated five times and the mean values of $\mathcal{L}_2[\widetilde{x}]$ and $E[\tau]$ calculated for each controller (see Table 5.2). The values confirm the visual analysis from the graphs, i.e., that the $\mathcal{H}_\infty$ control

Table 5.1 Values of γ obtained for LMI- and game theory-based $\mathcal{H}_\infty$ control of a three-link underactuated manipulator in configuration APA

Controller	Phase 1	Phase 2
Linear matrix inequalities	1.35	1.80
Game theory	1.90	1.90

Table 5.2 Performance indexes for LMI- and game theory-based $\mathcal{H}_\infty$ control of a three-link underactuated manipulator in configuration APA

Controller	$\mathcal{L}_2[\widetilde{x}]$	$E[\tau]$ (N m s)
Linear matrix inequalities	0.042	1.274
Game theory	0.049	0.929

via Linear Matrix Inequalities presents the best tracking performance, albeit at a higher energy cost.

5.6.4 Model-Based Adaptive Controller

In this and the next sections we validate the adaptive controllers presented in Sect. 5.5. Again we consider the underactuated manipulator in configuration APA. Due to the controllers' structure, however, in control phase 1 or APA$_u$,we use $q_c = [q_2]$ and $q_r = [q_1]$, i.e., the passive joint is controlled by applying torque to joint 1 while joint 3 is kept locked. In control phase 2, or APA$_l$, joint 2 is locked and the manipulator is controlled as if it were fully-actuated.

As we discussed in Sect. 5.3, $D(q,\dot{q})$ must be selected such that $\overline{N}(q,\dot{q})$ is skew-symmetric. In this case $\overline{M}(q)$ and $\overline{D}(q,\dot{q})$ are scalars given by:

$$\overline{M}(q) = M_{12}(q) - \frac{M_{11}(q)M_{22}(q)}{M_{21}(q)} \tag{5.24}$$

and

$$\overline{D}(q,\dot{q}) = D_{12}(q,\dot{q}) - \frac{M_{11}(q)D_{22}(q,\dot{q})}{M_{21}(q)}, \tag{5.25}$$

where $M_{ij}(q)$ and $D_{ij}(q,\dot{q})$ are the ij entries in matrices $M(q)$ and $D(q,\dot{q})$, respectively. $\overline{N}(q,\dot{q}) = \overline{D}(q,\dot{q}) - \frac{1}{2}\dot{\overline{M}}(q,\dot{q})$ is given by

$$\overline{N}(q,\dot{q}) = D_{12}(q,\dot{q}) - \frac{M_{11}(q)D_{22}(q,\dot{q})}{M_{21}(q)} - \frac{1}{2}\dot{M}_{12}(q,\dot{q})$$
$$+ \frac{1}{2}\frac{\dot{M}_{11}(q,\dot{q})M_{22}(q)}{M_{21}(q)} + \frac{1}{2}\frac{M_{11}(q)\dot{M}_{22}(q,\dot{q})}{M_{21}(q)}$$
$$- \frac{1}{2}\frac{M_{11}(q)M_{22}(q)\dot{M}_{21}(q,\dot{q})}{M_{21}^2(q)}. \quad (5.26)$$

Since $\overline{N}(q,\dot{q})$ is a scalar, it can only be skew-symmetric if it is equal to zero. Therefore D_{12} and D_{21} are selected such that the following conditions are satisfied:

$$D_{12} = \frac{1}{2}\dot{M}_{12}(q,\dot{q}) - \frac{1}{2}\frac{\dot{M}_{11}(q,\dot{q})M_{22}(q)}{M_{21}(q)}, \quad (5.27)$$

$$D_{22} = \frac{1}{2}\dot{M}_{22}(q,\dot{q}) - \frac{1}{2}\frac{M_{22}(q)\dot{M}_{21}(q,\dot{q})}{M_{21}(q)}. \quad (5.28)$$

The remaining entries in $D(q,\dot{q})$ can be computed from $V(q,\dot{q}) = D(q,\dot{q})\dot{q}$ as:

$$D_{11}(q,\dot{q}) = \frac{V_1(q,\dot{q}) - D_{12}(q,\dot{q})\dot{q}_{c2}}{\dot{q}_{c1}}, \quad D_{13} = 0,$$
$$D_{21}(q,\dot{q}) = \frac{V_2(q,\dot{q}) - D_{22}(q,\dot{q})\dot{q}_{c2}}{\dot{q}_{c1}}, \quad D_{23} = 0.$$

Recall that joint 3 is kept locked in control phase 1; therefore, the third line of $D(q,\dot{q})$ has no influence on the control system and, for convenience, can be selected as the third line of $C(q,\dot{q})$.

The best level of attenuation for the Model-based Adaptive Controller is $\gamma = 2$ for both control phases. The weighting matrices are $Q_1 = 40$, $Q_2 = 1$, $Q_{12} = 0$, and $R = 3$ in control phase 1 and $Q_1 = 3I_2$, $Q_2 = I_2$, $Q_{12} = 0$, and $R = 3I_2$ in control phase 2.

In the experiments shown below, the initial position and the desired final position are, respectively, $q(0) = [0° \; 0° \; 0°]^T$ and $q(T^1, T^2) = [30° \; 20° \; 10°]^T$, where $T^1 = 1.0$ s and $T^2 = [4.0 \; 2.0]$ s are the trajectory duration time for control phases 1 and 2, respectively. An external disturbance, τ_d, is introduced as:

$$\tau_d = \begin{bmatrix} \left(0.1e^{\frac{-(t-t_{d_1})^2}{2\mu_1^2}} + 0.08e^{\frac{-(t-t_{d_2})^2}{2\mu_1^2}}\right)\sin(7\pi t) \\ 0.05e^{\frac{-(t-t_{d_1})^2}{2\mu_2^2}}\sin(8\pi t) \\ 0.02e^{\frac{-(t-t_{d_2})^2}{2\mu_2^2}}\sin(8\pi t) \end{bmatrix},$$

where $t_{d_1} = 0.6$ s, $t_{d_2} = 2.5$ s, $\mu_1 = 0.4$, and $\mu_2 = 0.3$.

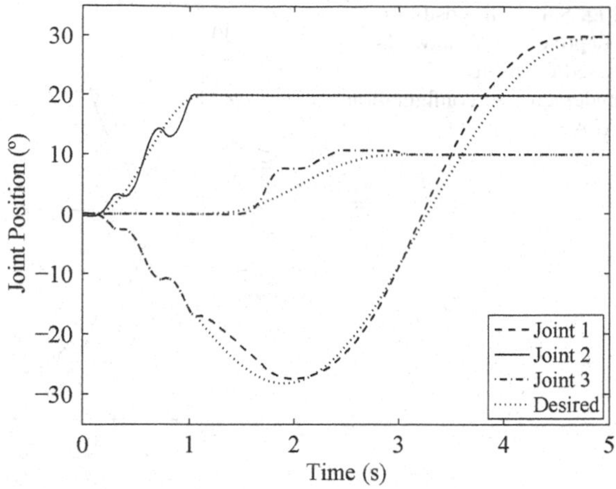

Fig. 5.6 Joint positions, adaptive model-based controller, underactuated configuration APA

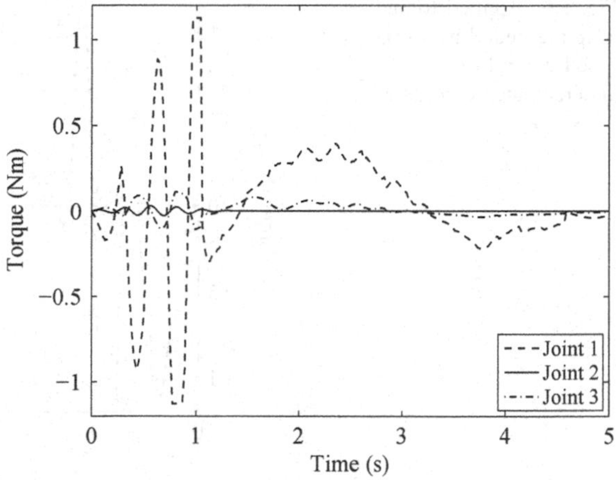

Fig. 5.7 Applied torques, adaptive model-based controller, underactuated configuration APA

The parameters $\overline{\theta}$ and θ and the regression matrices $\overline{Y}$ and Y for both control phases 1 and 2 are listed in the appendix to this chapter. The experimental results obtained with $\overline{\theta}(0) = [0\ldots0]^T_{7\times1}$ and $\overline{S} = 100$, and $\theta(0)$ given by the values of $\overline{\theta}$ at the end of phase 1 and $S = 100$, are shown in Figs. 5.6 and 5.7.

5.6.5 Neural Network-Based Adaptive Controller

For the adaptive neural network-based nonlinear $\mathcal{H}_\infty$ control, matrix $\overline{\Xi}$ in control phase 1 is selected as

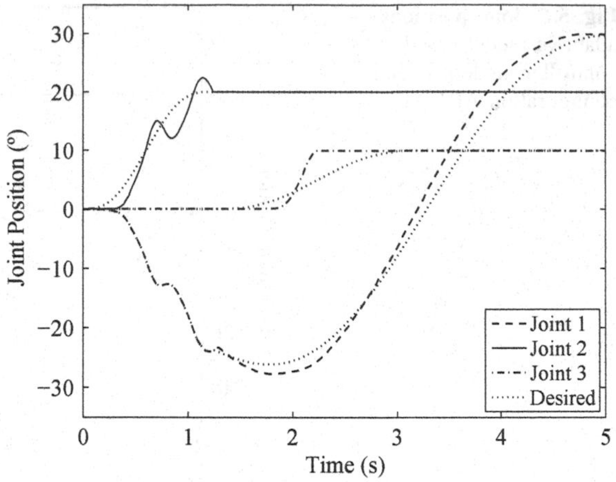

Fig. 5.8 Joint positions, adaptive neural network-based controller, underactuated configuration APA

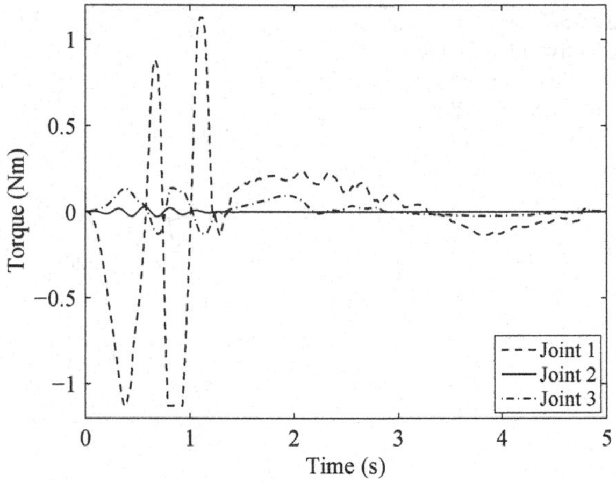

Fig. 5.9 Applied torques, adaptive neural network-based controller, underactuated configuration APA

$$\overline{\Xi} = [\,\xi_2\,]$$

and matrix Ξ in control phase 2 is selected as

$$\Xi = \begin{bmatrix} \xi_1 & 0 \\ 0 & \xi_3 \end{bmatrix}, \tag{5.29}$$

where $\xi_1 = [\xi_{11}, \ldots, \xi_{17}]$, $\xi_2 = [\xi_{21}, \ldots, \xi_{27}]$, $\xi_3 = [\xi_{31}, \ldots, \xi_{37}]$, with $\xi_{i1}, \ldots, \xi_{i7}$ (for $i = 1, 2, 3$) defined as for the fully-actuated case (Chap. 4). In control phase 1 the neural network parameter vector $\overline{\Theta}$ is

$$\overline{\Theta} = \left[\,\overline{\Theta}_{21}\ \overline{\Theta}_{22}\ \overline{\Theta}_{23}\ \overline{\Theta}_{24}\ \overline{\Theta}_{25}\ \overline{\Theta}_{26}\ \overline{\Theta}_{27}\,\right]^T,$$

Table 5.3 Performance indexes for adaptive model- and neural network-based $\mathcal{H}_\infty$ control of a three-link underactuated manipulator in configuration APA

Controller	$\mathcal{L}_2[\tilde{x}]$	$E[\tau]$ (N m s)
Model-based	0.0249	1.404
Neural network-based	0.0243	1.415

and in control phase 2 the parameter vector Θ is

$$\Theta = \begin{bmatrix} \Theta_1 \\ \Theta_3 \end{bmatrix},$$

with

$$\Theta_1 = [\Theta_{11}\ \Theta_{12}\ \Theta_{13}\ \Theta_{14}\ \Theta_{15}\ \Theta_{16}\ \Theta_{17}]^T,$$
$$\Theta_3 = [\Theta_{31}\ \Theta_{32}\ \Theta_{33}\ \Theta_{34}\ \Theta_{35}\ \Theta_{36}\ \Theta_{37}]^T.$$

The results for the adaptive neural network-based controller are shown in Figs. 5.8 and 5.9. As before, we conducted five experiments with each adaptive controller. Table 5.3 presents the mean values of $\mathcal{L}_2[\tilde{x}]$ and $E[\tau]$ for each one. The adaptive neural network-based nonlinear $\mathcal{H}_\infty$ controller presents a slightly better result with respect to the $\mathcal{L}_2$ performance index, but also presents greater energy spending.

Appendix

The variables used in the adaptive $\mathcal{H}_\infty$ controller are as follows:

- Parameter vector $\overline{\theta}$ for control phase 1: $\overline{\theta}_1 = \Delta(m_2 lc_2^2)$, $\overline{\theta}_2 = \Delta(m_3 l_2^2)$, $\overline{\theta}_3 = \Delta(m_3 lc_3^2)$, $\overline{\theta}_4 = \Delta(m_3 l_2 lc_3)$, $\overline{\theta}_5 = \Delta I_2$, $\overline{\theta}_6 = \Delta I_3$, and $\overline{\theta}_7 = \Delta f_2$.
- Matrix $\overline{Y}(\cdot)$ for control phase 1:

$$\overline{Y}(\cdot) = \left[\overline{Y}_{11}(\cdot)\overline{Y}_{12}(\cdot)\overline{Y}_{13}(\cdot)\overline{Y}_{14}(\cdot)\overline{Y}_{15}(\cdot)\overline{Y}_{16}(\cdot)\overline{Y}_{17}(\cdot)\right],$$

  where $\overline{Y}_{11}(\cdot) = y_{11}, \overline{Y}_{12}(\cdot) = y_{11}, \overline{Y}_{13}(\cdot) = y_{11}, \overline{Y}_{14}(\cdot) = 2\cos(q_3)y_{11} - \sin(q_3)\dot{q}_3 y_{21}, \overline{Y}_{15}(\cdot) = y_{11}, \overline{Y}_{16}(\cdot) = y_{11}, \overline{Y}_{17}(\cdot) = \dot{q}_2, y_{11} = \ddot{q}_2^d - T_{12}^{-1}T_{11}\dot{\tilde{q}}_2$, and $y_{21} = \dot{q}_2^d - T_{12}^{-1}T_{11}\tilde{q}_2$.
- Parameter vector $\overline{\theta}$ for control phase 2: $\theta_1 = \Delta(m_1 lc_1^2), \theta_2 = \Delta(m_2 l_1^2), \theta_3 = \Delta(m_2 l_1 lc_2), \theta_4 = \Delta(m_3 l_1^2), \theta_5 = \Delta(m_3 lc_3^2), \theta_6 = \Delta(m_3 l_1 lc_3), \theta_7 = \Delta I_1, \theta_8 = \Delta I_3, \theta_9 = \Delta f_1$, and $\theta_{10} = \Delta f_3$.

- Matrix $Y(\cdot)$ for control phase 2:

$$Y(\cdot)=\begin{bmatrix} Y_{11}(\cdot) & Y_{12}(\cdot) & Y_{13}(\cdot) & Y_{14}(\cdot) & Y_{15}(\cdot) & Y_{16}(\cdot) & Y_{17}(\cdot) & Y_{18}(\cdot) & Y_{19}(\cdot) & Y_{1,10}(\cdot) \\ Y_{21}(\cdot) & Y_{22}(\cdot) & Y_{23}(\cdot) & Y_{24}(\cdot) & Y_{25}(\cdot) & Y_{26}(\cdot) & Y_{27}(\cdot) & Y_{28}(\cdot) & Y_{29}(\cdot) & Y_{2,10}(\cdot) \end{bmatrix},$$

$$\begin{aligned}
Y_{11}(\cdot) &= 2y_{11}, \quad Y_{12}(\cdot) = y_{11},\\
Y_{13}(\cdot) &= 2\cos(q_2)y_{11} - \sin(q_2)\dot{q}_2y_{21} - \sin(q_2)(\dot{q}_1 + \dot{q}_2)y_{22},\\
Y_{14}(\cdot) &= (2 + 2\cos(q_2))y_{11} - \sin(q_2)\dot{q}_2y_{21},\\
Y_{15}(\cdot) &= y_{11} + y_{12},\\
Y_{16}(\cdot) &= (2\cos(q_2 + q_3) + 2\cos(q_3))y_{11}\\
&\quad + (\cos(q_2 + q_3) + \cos(q_3))y_{12} - \sin(q_2 + q_3)\dot{q}_2y_{21}\\
&\quad - (\sin(q_2 + q_3) + \sin(q_3))\dot{q}_3y_{21}\\
&\quad - (\sin(q_2 + q_3) + \sin(q_3))(\dot{q}_1 + \dot{q}_2 + \dot{q}_3)y_{22},\\
Y_{17}(\cdot) &= 2y_{11}, \quad Y_{18}(\cdot) = y_{11} + y_{12},\\
Y_{19}(\cdot) &= \dot{q}_1, \quad Y_{1,10} = 0,\\
Y_{21}(\cdot) &= 0, \quad Y_{22}(\cdot) = 0, \quad Y_{23}(\cdot) = 0,\\
Y_{24}(\cdot) &= 0, \quad Y_{25}(\cdot) = y_{11} + y_{12},\\
Y_{26}(\cdot) &= (\cos(q_2 + q_3) + \cos(q_3))y_{11}\\
&\quad + (\sin(q_2 + q_3) + \sin(q_3))\dot{q}_1y_{21} + \sin(q_3)\dot{q}_2y_{21},\\
Y_{27}(\cdot) &= 0, \; Y_{28}(\cdot) = y_{11} + y_{12}, \; Y_{29}(\cdot) = 0, \; Y_{2,10} = \dot{q}_2,
\end{aligned}$$

where $y_{11} = \ddot{q}_1^d - T_{12}^{-1}T_{11}\dot{\tilde{q}}_1$, $y_{12} = \ddot{q}_3^d - T_{12}^{-1}T_{11}\dot{\tilde{q}}_3$, $y_{21} = \dot{q}_1^d - T_{12}^{-1}T_{11}\tilde{q}_1$, and $y_{22} = \dot{q}_3^d - T_{12}^{-1}T_{11}\tilde{q}_3$.

References

1. Arai H, Tanie K, Tachi S (1991) Dynamic control of a manipulator with passive joints in operational space. IEEE Trans Rob Autom 9(1):85–93
2. Bergerman M (1996) Dynamics and control of underactuated manipulators. Ph.D. Thesis, Carnegie Mellon University, Pittsburgh, p 129
3. Visinsky ML, Cavallaro JR, Walker ID (1994) Robotic fault detection and fault tolerance: a survey. Reliab Eng Syst Saf 46(2):139–158
4. Visinsky ML, Cavallaro JR, Walker ID (1995) A dynamic fault tolerance framework for remote robots. IEEE Trans Rob Autom 11(4):477–490
5. Siqueira AAG, Terra MH (2004) Nonlinear and Markovian $\mathcal{H}_\infty$ controls of underactuated manipulators. IEEE Trans Control Syst Technol 12(6):811–826
6. Terra MH, Tinós R (2001) Fault detection and isolation in robotic manipulators via neural networks: a comparison among three architectures for residual analysis. J Rob Syst 18(7):357–374

7. Dixon WE, Walker ID, Dawson DM, Hartranft JP (2000) Fault detection for robot manipulators with parametric uncertainty: a prediction-error-based approach. IEEE Trans Rob Autom 16(6):689–699
8. Routh EJ (1960) Dynamics of a system of rigid bodies, 7th edn. Dover Publications Inc., New York
9. Wu F, Yang XH, Packard A, Becker G (1996) Induced $\mathcal{L}_2$-norm control for LPV systems with bounded parameter variation rates. Int J Rob Nonlinear Control 6(9–10):983–998
10. Apkarian P, Adams RJ (1998) Advanced gain-scheduling techniques for uncertain systems. IEEE Trans Control Syst Technol 6(1):21–32

Chapter 6
Markov Jump Linear Systems-Based Control

6.1 Introduction

A large number of dynamic systems are inherently vulnerable to abrupt changes in their structures caused by, for example, component failures, sudden environmental disturbances, and abrupt variation of the operating point of a nonlinear plant. This class of systems can be modeled by a set of linear systems with transitions among models determined by a Markov chain taking values in a finite set. An important point in this process is to develop Markovian jump models with probability matrices of Markovian state transitions.

In this chapter, we present fault tolerant systems based on Markovian control theory for three-link manipulator robots. We present procedures to model the changes in the operating points of the system and to model the probability of a fault occurrence. Both models are grouped in a single Markovian jump model. We assume that the fault is detected and isolated with the filtered torque prediction error approach proposed in [9].

The framework presented consists of a complete Markovian jump model of a three-link manipulator, incorporating all possible fault combinations. Additionally, we compare the performance of four Markovian controllers, $\mathcal{H}_2$, $\mathcal{H}_\infty$, mixed $\mathcal{H}_2/\mathcal{H}_\infty$ based on state-feedback [4, 6, 7], and a $\mathcal{H}_\infty$ output feedback-based approach [8]. We present two sets of experiments, one where only one fault occurs and one where two faults occur in sequence. The results presented in this chapter show that it is possible to accommodate a sequence of abrupt changes in the manipulator without the necessity of stopping it completely to modify the control strategy after a fault occurs. This is in contrast with our earlier work where we showed that, when the manipulator is moving, deterministic controllers cannot guarantee stability after a fault [11]. Those experiments are reproduced here using

A. A. G. Siqueira et al., *Robust Control of Robots*,
DOI: 10.1007/978-0-85729-898-0_6,

the computed torque plus linear $\mathcal{H}_\infty$ controllers for robotic manipulators with free joint faults.

This chapter is organized as follows: Sect. 6.2 motivates the development of the Markovian jump model through a working example; Sect. 6.3 presents the complete Markovian jump model for a three-link robotic manipulator; Sect. 6.4 presents the equations used to compute the $\mathcal{H}_2$, $\mathcal{H}_\infty$, and mixed $\mathcal{H}_2/\mathcal{H}_\infty$ controllers; Sect. 6.5 presents an $\mathcal{H}_\infty$ output feedback control approach for Markovian jump linear systems; and Sect. 6.6 presents the results of the these Markovian controllers applied to the UARM.

6.2 Motivation

To motivate the Markovian controllers developed in this chapter, we initially try to control the three-link manipulator using a mixed computed torque plus linear $\mathcal{H}_\infty$ deterministic control strategy, described in detail in Chap. 2. The manipulator starts operation in fully-actuated mode with initial position $q(0) = [0° \, 0° \, 0°]^T$ and desired final position $q(T) = [20° \, 20° \, 20°]^T$, where $T = [4.0 \; 4.0 \; 4.0]$ s is the duration of the motion. When all joints reach approximately 15° at $t_f = 2.2$ s, we introduce an artificial free joint fault in the second joint by disabling the joint's actuator. We assume that fault detection is perfect and instantaneous.

The deterministic control strategy is applied before and after the fault occurrence. If the robot stops completely when the fault is detected and before the post-fault control strategy is applied, the stability is guaranteed and the robot reaches the desired position. On the other hand, if the robot continues moving through the fault, the controller cannot guarantee the stability of the system (see Figs. 6.1 and 6.2). As we will show at the end of this chapter, the Markovian controller is able to

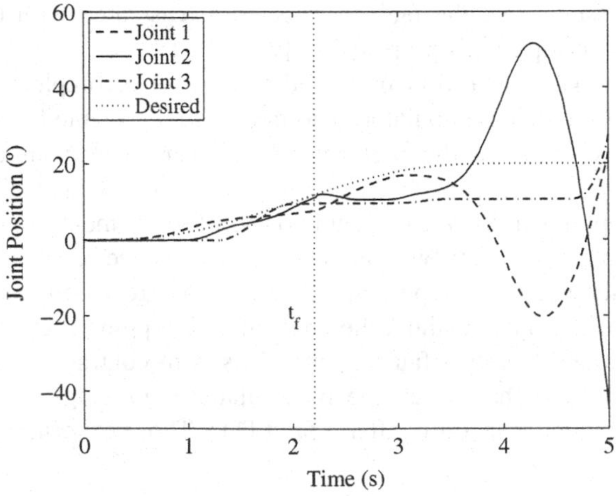

Fig. 6.1 Joint positions, computed torque plus linear $\mathcal{H}_\infty$ control with fault occurrence (t_f = fault time)

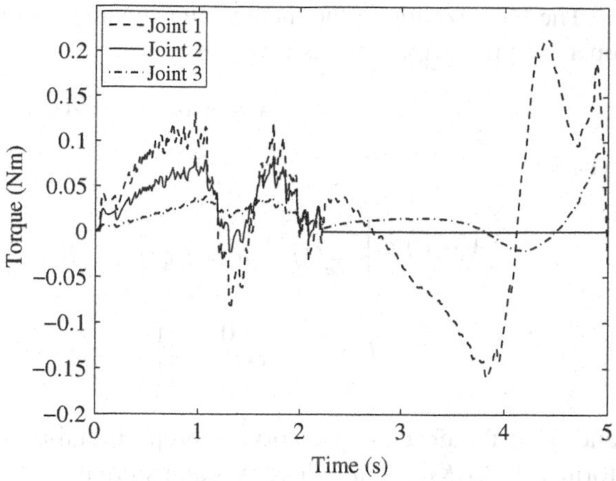

Fig. 6.2 Applied torques, computed torque plus linear $\mathcal{H}_\infty$ control with fault occurrence

bring the robot to its desired position without interruption in the motion even in the presence of multiple faults.

6.3 Markovian Jump Linear Model

We start with the dynamic model of the fully-actuated manipulator with n joints

$$\tau = M(q)\ddot{q} + b(q, \dot{q}). \tag{6.1}$$

When free joints are present, the dynamic model can be represented as in Eq. 5.17:

$$\begin{bmatrix} \tau_a \\ 0 \end{bmatrix} = \begin{bmatrix} M_{ac}(q) & M_{ar}(q) \\ M_{uc}(q) & M_{ur}(q) \end{bmatrix} \begin{bmatrix} \ddot{q}_c \\ \ddot{q}_r \end{bmatrix} + \begin{bmatrix} b_a(q,\dot{q}) \\ b_u(q,\dot{q}) \end{bmatrix} \tag{6.2}$$

(the locked passive joints do not contribute to the motion of the system, and therefore are eliminated from the model). Isolating the vector $\ddot{q}_r$ in the second line of (6.2) and substituting in the first line, we obtain the

Active joints torque vector:

$$\tau_a = \overline{M}(q)\ddot{q}_c + \overline{b}(q, \dot{q}), \tag{6.3}$$

where

$$\overline{M}(q) = M_{ac}(q) - M_{ar}(q)M_{ur}^{-1}(q)M_{uc}(q),$$
$$\overline{b}(q, \dot{q}) = b_a(q, \dot{q}) - M_{ar}(q)M_{ur}^{-1}(q)b_u(q, \dot{q}).$$

The linearization of the manipulator dynamics, represented by (6.3), around an operating point $(q_0, \dot{q}_0)$, is given by:

$$\dot{x} = \bar{A}(q,\dot{q})x + \bar{B}(q)\tau,$$

where

$$\bar{A}(q,\dot{q}) = \left.\begin{bmatrix} 0 & I \\ -\frac{\partial}{\partial q}(\overline{M}^{-1}(q)\bar{b}(q,\dot{q})) & -\overline{M}^{-1}(q)\frac{\partial}{\partial \dot{q}}(\bar{b}(q,\dot{q})) \end{bmatrix}\right|_{(q_0,\dot{q}_0)},$$

$$\bar{B}(q) = \left.\begin{bmatrix} 0 \\ \overline{M}^{-1}(q) \end{bmatrix}\right|_{(q_0)}, \quad x = \begin{bmatrix} q^d - q \\ \dot{q}^d - \dot{q} \end{bmatrix},$$

and q^d is the desired trajectory. A proportional-derivative (PD) controller of the form $\tau = [K_P \, K_D]x + u$ (where u is an external control input) can be introduced to pre-compensate model uncertainty. Hence, the dynamic equation of the manipulator can be written as:

Extended linearized system:

$$\begin{aligned} \dot{x} &= \widetilde{A}(q,\dot{q})x + \widetilde{B}(q)u, \\ z &= \widetilde{C}x + \widetilde{D}u, \end{aligned} \tag{6.4}$$

where

$$\tilde{A}(q,\dot{q}) = \left.\begin{bmatrix} 0 & I \\ -\frac{\partial}{\partial q}(\bar{M}^{-1}(q)\bar{b}(q,\dot{q})) + \bar{M}^{-1}(q)K_P & -\bar{M}^{-1}(q)\left[\frac{\partial}{\partial \dot{q}}(\bar{b}(q,\dot{q})) - K_D\right] \end{bmatrix}\right|_{\bar{q}_0},$$

$$\widetilde{B}(q) = \bar{B}(q), \quad \widetilde{C} = \begin{bmatrix} \alpha I & 0 \\ 0 & 0 \end{bmatrix},$$

$$\widetilde{D} = \begin{bmatrix} 0 \\ \beta I \end{bmatrix}, \quad \tau = [K_P \quad K_D]x + u,$$

where $\bar{q}_0 = (q_0, \dot{q}_0)$, z is the output variable of the manipulator and α and β are constants defined by the designer to adjust the Markovian controllers.

The state feedback $\mathcal{H}_2$, $\mathcal{H}_\infty$ and mixed $\mathcal{H}_2/\mathcal{H}_\infty$ controllers presented in this chapter were developed for discrete time systems. Therefore, we discretize (6.4) as:

$$\begin{aligned} x(k+1) &= A(q,\dot{q})x(k) + B(q)\tau(k), \\ z(k) &= Cx(k) + D\tau(k). \end{aligned} \tag{6.5}$$

Table 6.1 Controlled joints in configurations AAP, APA, and PAA

Configuration	Control phase 1	Control phase 2
AAP	1, 3	1, 2
APA	2, 3	1, 3
PAA	1, 3	2, 3

The Markovian jump model developed in this section describes the changes in the linearization points of the plant (6.5), and the probability of a fault occurrence for a three-link manipulator [10, 12]. Although the changes in the system linearization points are not genuine stochastic events, in contrast with a fault occurrence, the Markovian techniques can be applied in this case since the jump probability is related with the expected mean time the system is supposed to lie in each state of the Markovian chain.

6.3.1 Configuration After Fault Occurrence

For a 3-link manipulator robot, seven possible fault configurations can occur: AAP, APA, PAA, APP, PAP, PPA, and PPP, where A means that the corresponding joint is active and P means that it is passive. For example, in the AAP configuration joints 1 and 2 are active and joint 3 is passive.

In the faulty configurations AAP, APA, and PAA, $n_a = 2$; therefore, two control phases are necessary to control all joints to the set-point (see Chap. 5 and [1, 2, 3] for details). In the first control phase, the vector of controlled joints, q_c, contains the passive joint and one active joint. In the second control phase, q_c contains the active joints and the passive joint is kept locked since it has already reached the set-point. Table 6.1 summarizes the joints controlled in the two control phases for each configuration.

The first control phase is denoted by the configuration name followed by the subscript u (to indicate that the passive joint is unlocked); and the second phase by the subscript l (to indicate that the passive joint is locked). For example, APA_u and APA_l represent the first and the second control phases of the configuration APA, respectively.

In the faulty configurations APP, PAP, and PPA, $n_a = 1$; therefore three control phases are necessary to control all joints to the set-point. In the first control phase, the vector of controlled joints, q_c, contains one passive joint. In the second control phase, the other passive joint is selected to form the vector of controlled joints. In the last control phase, the active joint is controlled. The passive joints not being controlled in each control phase are kept locked. Table 6.2 summarizes the joints controlled in the three control phases for each configuration.

The first control phase is denoted by the configuration name followed by the subscript $u1$; the second control phase by the subscript $u2$; and the third control phase by the subscript l. For example, PPA_{u1}, PPA_{u2}, and PPA_l represent the first, the second and the third control phases of the configuration PPA, respectively.

Table 6.2 Controlled joint in configurations APP, PAP, and PPA

Configuration	Control phase 1	Control phase 2	Control phase 3
APP	2	3	1
PAP	1	3	2
PPA	1	2	3

Table 6.3 AAA–APA Markovian states and linearization points

Markovian states			Linearization Points					
AAA	APA_u	APA_l	q_1	q_2	q_3	$\dot{q}_1$	$\dot{q}_2$	$\dot{q}_3$
1	9	17	5	5	5	0	0	0
2	10	18	15	5	5	0	0	0
3	11	19	5	15	5	0	0	0
4	12	20	15	15	5	0	0	0
5	13	21	5	5	15	0	0	0
6	14	22	15	5	15	0	0	0
7	15	23	5	15	15	0	0	0
8	16	24	15	15	15	0	0	0

6.3.2 Linearization Points

To define the linearization points, the workspace of each joint is divided in sectors, denoted *sec*. For each combination of *sec*/2 for each joint, we define a linearization point for the manipulator. The choice of these sectors and the number of sectors, n_{sec}, needs to be done in order to guarantee the effectiveness of the Markovian jump model. In the experiments shown in this chapter, the workspace of each joint is divided in two sectors, with $sec = 10°$ (the set-point is defined as 20° for each joint, with initial position 0°). We define a central point for each sector at 5° for the first sector and 15° for the second one. All the possible combinations of positioning of the three joints, q_1, q_2, q_3, at these two points are used to map the manipulator workspace. Setting the velocities to zero, we end up with eight linearization points, shown in Table 6.3.

6.3.3 Markovian States

The Markovian states are the manipulator's discrete dynamic model (6.5) linearized around the eight points for all control phases of all configurations. Recall that configuration AAA has one control phase, faulty configurations AAP, APA, and PAA have each two control phases (for a total of six), faulty configurations PPA, PAP, and PPA have each three control phases (for a total of nine), and configuration PPP represents only one state, independently of how many linearization points are used. Therefore the Markovian jump model has $8 \times (1 + 6 + 9) + 1 = 129$ states. Accordingly, the number of Markovian states for an n-link manipulator, T_{MS}, is given by:

$$T_{MS} = 1 + n_{lp} \times \left(\sum_{i=1}^{n-1} (n_{cp_i} \times n_{fc_i}) + 1 \right), \tag{6.6}$$

where $n_{fc_i} = \frac{n!}{i!(n-i)!}$ is the number of possible fault configurations for i faults, $n_{cp_i} = ceil(\frac{n}{n-i})$ is the number of control phases for a configuration with i faults ($ceil(x)$ rounds x to the nearest integer towards infinity) and $n_{lp} = (n_{sec})^n$ is the number of linearization points.

Figure 6.3 presents the complete Markovian jump model, describing all possible fault occurrences for a 3-link robotic manipulator. The probability matrices P_f, P_s, P_0 and P_{100} indicate, respectively, the probability of a fault occurrence, the probability that the passive joint is controlled to reach the set-point, the probability that a free joint is repaired (here $P_0 = 0$), and the probability that the manipulator is in the configuration PPP (since $P_0 = 0$, $P_{100} = 1$).

6.3.4 AAA–APA Fault Sequence

The AAA–APA fault sequence is represented in the Markovian jump model by the numbers 1, 2, and 3 in Fig. 6.3. The system starts in the configuration AAA. When a fault occurs, the system goes to control phase APA$_u$ at the same linearization point. When the second joint reaches the set-point, the system goes to control phase APA$_l$.

According to Table 6.1, the vector of controlled joints, q_c, is chosen as $q_c = [q_2\ q_3]^T$ for control phase APA$_u$, and $q_c = [q_1\ q_3]^T$ for control phase APA$_l$. For each linearization point there exist three sets of matrices $A(q, \dot{q})$, $B(q)$, C, and D in (6.5), corresponding to the control phases AAA, APA$_u$, and APA$_l$. The dimensions of these matrices for configurations APA$_u$ and APA$_l$ are smaller than for configuration AAA. However, for the Markovian controllers adopted here, all linear systems must have the same dimension. To solve this problem, rows and columns of zeros are added to the matrices.

According to Eq. 6.6 there exist 24 Markovian states for this fault sequence (see Table 6.3). Following the Markovian theory it is necessary to group them in a transition probability matrix P of dimension 24×24. The element p_{ij} of P represents the probability of the system, being in the Markovian state i, to go to the state j at the next time k. This implies in $\sum_j p_{ij} = 1$ for any line i of P.

Matrix P is partitioned into nine 8×8 submatrices, shown in Eq. 6.7. The elements of P were selected empirically. The submatrix P_{AAA} groups the relations between linearization points of normal operation in the configuration AAA, and the diagonal submatrix P_f groups the probabilities of a fault occurrence when the system is in normal operation. P_f in the first line of P represents that when a fault occurs, the system goes from the configuration AAA to the control phase APA$_u$. After the fault, the system will be in the control phase APA$_u$, and the system

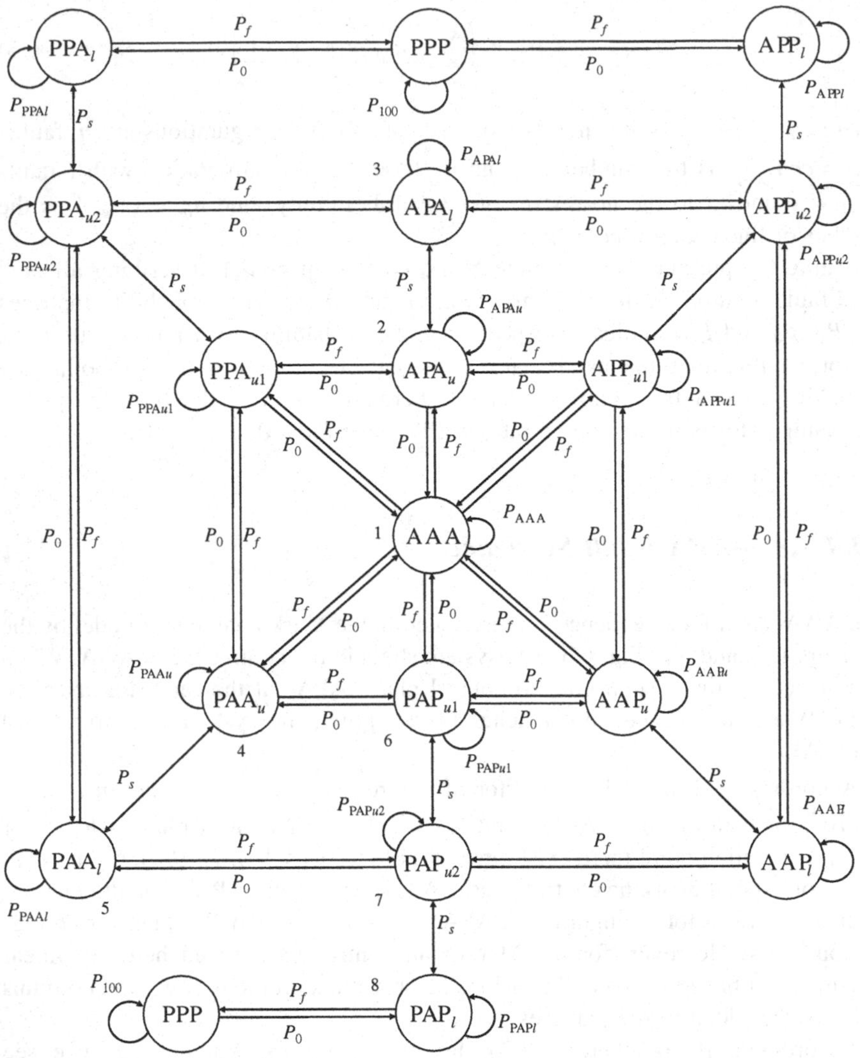

Fig. 6.3 Markovian jump model for the 3-link manipulator UARM

changes to the second line of P, where $P_{\mathrm{APA_u}}$ groups the relations between the linearization points in the control phase APA$_u$. $P_0 = 0$ represents the fact that the free joint cannot be repaired, and P_s represents the probability that the system will go to the control phase APA$_l$. After that, the system can be in control phases APA$_u$ or APA$_l$, according to P_s. In the third line of P, $P_{\mathrm{APA_l}}$ groups the relations between the linearization points in the set APA$_l$, P_s represents the probability of the system to return to the control phase APA$_u$, and P_0 represents, again, the impossibility of repairing the free joint.

$$P = \left[\begin{array}{c|c|c} P_{\mathrm{AAA}} & P_f & P_0 \\ \hline P_0 & P_{\mathrm{APA_u}} & P_s \\ \hline P_0 & P_s & P_{\mathrm{APA_l}} \end{array}\right], \tag{6.7}$$

with

$$P_{AAA} = \begin{bmatrix} 0.89 & 0.10 & 0 & 0 & 0 & 0 & 0 & 0 \\ 0.10 & 0.79 & 0.10 & 0 & 0 & 0 & 0 & 0 \\ 0 & 0.10 & 0.79 & 0.10 & 0 & 0 & 0 & 0 \\ 0 & 0 & 0.10 & 0.79 & 0.10 & 0 & 0 & 0 \\ 0 & 0 & 0 & 0.10 & 0.79 & 0.10 & 0 & 0 \\ 0 & 0 & 0 & 0 & 0.10 & 0.79 & 0.10 & 0 \\ 0 & 0 & 0 & 0 & 0 & 0.10 & 0.79 & 0.10 \\ 0 & 0 & 0 & 0 & 0 & 0 & 0.10 & 0.89 \end{bmatrix},$$

$$P_f = \begin{bmatrix} 0.01 & 0 & 0 & 0 & 0 & 0 & 0 & 0 \\ 0 & 0.01 & 0 & 0 & 0 & 0 & 0 & 0 \\ 0 & 0 & 0.01 & 0 & 0 & 0 & 0 & 0 \\ 0 & 0 & 0 & 0.01 & 0 & 0 & 0 & 0 \\ 0 & 0 & 0 & 0 & 0.01 & 0 & 0 & 0 \\ 0 & 0 & 0 & 0 & 0 & 0.01 & 0 & 0 \\ 0 & 0 & 0 & 0 & 0 & 0 & 0.01 & 0 \\ 0 & 0 & 0 & 0 & 0 & 0 & 0 & 0.01 \end{bmatrix},$$

$$P_0 = \begin{bmatrix} 0 & 0 & 0 & 0 & 0 & 0 & 0 & 0 \\ 0 & 0 & 0 & 0 & 0 & 0 & 0 & 0 \\ 0 & 0 & 0 & 0 & 0 & 0 & 0 & 0 \\ 0 & 0 & 0 & 0 & 0 & 0 & 0 & 0 \\ 0 & 0 & 0 & 0 & 0 & 0 & 0 & 0 \\ 0 & 0 & 0 & 0 & 0 & 0 & 0 & 0 \\ 0 & 0 & 0 & 0 & 0 & 0 & 0 & 0 \\ 0 & 0 & 0 & 0 & 0 & 0 & 0 & 0 \end{bmatrix},$$

$$P_{\mathrm{APA_u}} = \begin{bmatrix} 0.78 & 0.20 & 0 & 0 & 0 & 0 & 0 & 0 \\ 0.10 & 0.78 & 0.10 & 0 & 0 & 0 & 0 & 0 \\ 0 & 0.10 & 0.78 & 0.10 & 0 & 0 & 0 & 0 \\ 0 & 0 & 0.10 & 0.78 & 0.10 & 0 & 0 & 0 \\ 0 & 0 & 0 & 0.10 & 0.78 & 0.10 & 0 & 0 \\ 0 & 0 & 0 & 0 & 0.10 & 0.78 & 0.10 & 0 \\ 0 & 0 & 0 & 0 & 0 & 0.10 & 0.78 & 0.10 \\ 0 & 0 & 0 & 0 & 0 & 0 & 0.20 & 0.78 \end{bmatrix},$$

$$P_{\mathrm{APA}_l} = P_{\mathrm{APA}_u}, \quad \text{and} \quad P_s = 2P_f.$$

Table 6.4 AAA–PAA–PAP Markovian states

AAA	PAA_u	PAA_l	PAP_{u1}	PAP_{u2}	PAP_l
1	9	17	25	33	41
2	10	18	26	34	42
3	11	19	27	35	43
4	12	20	28	36	44
5	13	21	29	37	45
6	14	22	30	38	46
7	15	23	31	39	47
8	16	24	32	40	48

6.3.5 AAA–PAA–PAP Fault Sequence

The AAA–PAA–PAP fault sequence is represented in the Markovian jump model, Fig. 6.3, by the numbers 1, 4, 5, 6, 7, and 8. The Markovian states for this fault sequence are shown in Table 6.4, with the same linearization points adopted in the AAA–APA fault sequence (see Table 6.3). The system starts in configuration AAA. If a fault occurs in joint 1, the system maintains the linearization point and goes to the corresponding point in control phase PAA_u. If the second fault (joint 3) occurs during control phase PAA_u, the system goes to control phase PAP_{u1}. If the faults in joints 1 and 3 occur at the same time, the system goes from configuration AAA to control phase PAP_{u1}. The second fault can also occur during control phase PAA_l, when passive joint 1 has already reached the set-point .

According to Tables 6.1 and 6.2, the vector of controlled joints, q_c, is chosen as $q_c = [q_1\ q_3]^T$ for control phase PAA_u; $q_c = [q_2\ q_3]^T$ for control phase PAA_l; $q_c = q_1$ for PAP_{u1}; $q_c = q_3$ for PAP_{u2}; and $q_c = q_2$ for PAP_l. In this case, for each linearization point there exist six sets of matrices $A(q, \dot{q})$, $B(q)$, C, and D in (6.5), corresponding to the control phases AAA, PAA_u, PAA_l, PAP_{u1}, PAP_{u2}, and PAP_l. Again, rows and columns of zeros must be added to these matrices to guarantee that all linear systems have the same dimension.

For this fault sequence, there exist $T_{\text{MS}} = 48$ Markovian states; therefore the transition probability matrix P must be of dimension 48×48. Analogously to the AAA–APA fault sequence, matrix P is partitioned in 36 submatrices of dimension 8×8, shown in Eq. 6.8. The submatrices P_0, P_f, and P_s are the same used in Sect. 6.3.4 for the AAA–APA fault sequence.

$$P = \left[\begin{array}{c|c|c|c|c|c} P_{\text{AAA}} & P_f & P_0 & P_f & P_0 & P_0 \\ \hline P_0 & P_{\text{PAA}_u} & P_s & P_f & P_0 & P_0 \\ \hline P_0 & P_s & P_{\text{PAA}_l} & P_0 & P_f & P_0 \\ \hline P_0 & P_0 & P_0 & P_{\text{PAP}_{u1}} & P_s & P_0 \\ \hline P_0 & P_0 & P_0 & P_s & P_{\text{PAP}_{u2}} & P_s \\ \hline P_0 & P_0 & P_0 & P_s & P_s & P_{\text{PAP}_l} \end{array}\right], \tag{6.8}$$

with

$$P_{\mathrm{AAA}} = \begin{bmatrix} 0.88 & 0.10 & 0 & 0 & 0 & 0 & 0 & 0 \\ 0.10 & 0.78 & 0.10 & 0 & 0 & 0 & 0 & 0 \\ 0 & 0.10 & 0.78 & 0.10 & 0 & 0 & 0 & 0 \\ 0 & 0 & 0.10 & 0.78 & 0.10 & 0 & 0 & 0 \\ 0 & 0 & 0 & 0.10 & 0.78 & 0.10 & 0 & 0 \\ 0 & 0 & 0 & 0 & 0.10 & 0.78 & 0.10 & 0 \\ 0 & 0 & 0 & 0 & 0 & 0.10 & 0.78 & 0.10 \\ 0 & 0 & 0 & 0 & 0 & 0 & 0.10 & 0.88 \end{bmatrix},$$

$$P_{\mathrm{PAA}_u} = \begin{bmatrix} 0.77 & 0.20 & 0 & 0 & 0 & 0 & 0 & 0 \\ 0.10 & 0.77 & 0.10 & 0 & 0 & 0 & 0 & 0 \\ 0 & 0.10 & 0.77 & 0.10 & 0 & 0 & 0 & 0 \\ 0 & 0 & 0.10 & 0.77 & 0.10 & 0 & 0 & 0 \\ 0 & 0 & 0 & 0.10 & 0.77 & 0.10 & 0 & 0 \\ 0 & 0 & 0 & 0 & 0.10 & 0.77 & 0.10 & 0 \\ 0 & 0 & 0 & 0 & 0 & 0.10 & 0.77 & 0.10 \\ 0 & 0 & 0 & 0 & 0 & 0 & 0.20 & 0.77 \end{bmatrix},$$

$$P_{\mathrm{PAP}_{u1}} = \begin{bmatrix} 0.78 & 0.20 & 0 & 0 & 0 & 0 & 0 & 0 \\ 0.10 & 0.78 & 0.10 & 0 & 0 & 0 & 0 & 0 \\ 0 & 0.10 & 0.78 & 0.10 & 0 & 0 & 0 & 0 \\ 0 & 0 & 0.10 & 0.78 & 0.10 & 0 & 0 & 0 \\ 0 & 0 & 0 & 0.10 & 0.78 & 0.10 & 0 & 0 \\ 0 & 0 & 0 & 0 & 0.10 & 0.78 & 0.10 & 0 \\ 0 & 0 & 0 & 0 & 0 & 0.10 & 0.78 & 0.10 \\ 0 & 0 & 0 & 0 & 0 & 0 & 0.20 & 0.78 \end{bmatrix},$$

$$P_{\mathrm{PAP}_{u2}} = \begin{bmatrix} 0.76 & 0.20 & 0 & 0 & 0 & 0 & 0 & 0 \\ 0.10 & 0.76 & 0.10 & 0 & 0 & 0 & 0 & 0 \\ 0 & 0.10 & 0.76 & 0.10 & 0 & 0 & 0 & 0 \\ 0 & 0 & 0.10 & 0.76 & 0.10 & 0 & 0 & 0 \\ 0 & 0 & 0 & 0.10 & 0.76 & 0.10 & 0 & 0 \\ 0 & 0 & 0 & 0 & 0.10 & 0.76 & 0.10 & 0 \\ 0 & 0 & 0 & 0 & 0 & 0.10 & 0.76 & 0.10 \\ 0 & 0 & 0 & 0 & 0 & 0 & 0.20 & 0.76 \end{bmatrix},$$

6.4 MJLS Robust Control Based on State Feedback

We now turn to the design and comparison of three controllers based on the Markovian jump model developed in Sect. 6.3. They are the $\mathcal{H}_2$, $\mathcal{H}_\infty$, and mixed $\mathcal{H}_2/\mathcal{H}_\infty$ Markovian controllers proposed in [4–7]. We start with the discrete linear system

$$\begin{aligned} x(k+1) &= A_{\Theta(k)}x(k) + B_{\Theta(k)}u(k) + W_{\Theta(k)}w(k), \\ z(k) &= C_{\Theta(k)}x(k) + D_{\Theta(k)}u(k), \\ x(0) &= x_0, \qquad \Theta(0) = \Theta_0, \end{aligned} \tag{6.9}$$

subject to Markovian jumps, where $A_{\Theta(k)} = (A_1, \ldots, A_N) \in \mathbb{H}^n$, $B_{\Theta(k)} = (B_1, \ldots, B_N) \in \mathbb{H}^{m,n}$, $W_{\Theta(k)} = (W_1, \ldots, W_N) \in \mathbb{H}^{r,n}$, $w = (w(0), w(1), \ldots) \in l_2^r$, $C_{\Theta(k)} = (C_1, \ldots, C_N) \in \mathbb{H}^{n,s}$, and $D_{\Theta(k)} = (D_1, \ldots, D_N) \in \mathbb{H}^{m,s}$ with $D_i^* D_i > 0$ for all i.

$\Theta(k)$ is a Markov chain with values in $\{1, \ldots, N\}$, and Θ_0 is its initial condition. Conjugate transpose is denoted by *, $\mathbb{H}^{m,n}$ ($\mathbb{H}^{n,n} = \mathbb{H}^n$) is a linear space made up of all sequence of complex matrices, and l_2^r is the Hilbert set of random variables of second order $w = \big(w(0), w(1), \ldots\big)$ with $w(k) \in \mathbb{R}^r$, where

$$\|w\|_2^2 = \sum_{k=0}^{\infty} \|w(k)\|_2^2 < \infty, \quad \text{and} \quad \|w(k)\|_2^2 = E(\|w(k)\|^2).$$

We define the operator $\mathcal{E}(.) = \big(\mathcal{E}_1(.), \ldots, \mathcal{E}_N(.)\big) \in \mathbb{B}(\mathbb{R}^n)$ and for all sequence of complex matrices $X = (X_1, \ldots, X_N) \in \mathbb{H}^n$ (with $X_i \in \mathbb{B}(\mathbb{C}^m, \mathbb{C}^n)$ for $i = 1, \ldots, N$),

$$\mathcal{E}_i(X) = \sum_{j=1}^{N} p_{ij} X_j,$$

where $\mathbb{B}(\mathbb{C}^m, \mathbb{C}^n)$ denotes the normed linear space of all $n \times m$ complex matrices and $\mathbb{B}(\mathbb{R}^m, \mathbb{R}^n)$ denotes the normed linear space of all $n \times m$ real matrices ($\mathbb{B}(\mathbb{C}^n)$ and $\mathbb{B}(\mathbb{R}^n)$ whenever $n = m$). Note that the system (6.9) represents a collection of all linearized systems (6.5) which are stochastically chosen to represent the manipulator dynamics at each instant. For the purposes of this chapter, the purpose of this Markovian system is to describe scenarios of possible faults which can occur during the manipulator operation. Note also that we are adding in this system possible disturbances $w(k)$ which can affect the positions, velocities, and accelerations of the manipulator. In the examples presented in this chapter, disturbances are applied in the joint torques of the manipulator in order to test the robustness of the controllers.

6.4.1 $\mathcal{H}_2$ Control

The $\mathcal{H}_2$ control technique aims to minimize the quadratic functional:

$$J(x,\Theta,u) = \frac{1}{2}\sum_{k=0}^{\infty} E\left\{\begin{bmatrix} x^*(k) & u^*(k)\end{bmatrix}\begin{bmatrix} Q_{\Theta(k)} & L_{\Theta(k)} \\ L^*_{\Theta(k)} & R_{\Theta(k)}\end{bmatrix}\begin{bmatrix} x(k) \\ u(k)\end{bmatrix}\right\}, \tag{6.10}$$

constrained by (6.9) with $w = 0$. $L_{\Theta(k)} = (L_1,\ldots,L_N) \in \mathbb{H}^{m,n}$, $Q_{\Theta(k)} = (Q_1,\ldots,Q_N) \in \mathbb{H}^{n^*}$, and $R_{\Theta(k)} = (R_1,\ldots,R_N) \in \mathbb{H}^{m^*}$. All matrices Q_i and R_i $(i = 1,\ldots,N)$ should be Hermitian.

$\mathcal{H}_2$ control law: The control law that minimizes the functional (6.10) is given by

$$u(k) = F_{\Theta(k)}x(k),$$

where $F_{\Theta(k)} = (F_1,\ldots,F_N)$ and

$$F_i = -\left(B_i^*\mathcal{E}_i(X)B_i + R_i\right)^{-1}\left(B_i^*\mathcal{E}_i(X)A_i + L_i^*\right),$$

X is the solution of the following coupled algebraic Riccati equations [7]

$$\begin{aligned} 0 = & -X_i + A_i^*\mathcal{E}_i(X)A_i + Q_i \\ & - \left(A_i^*\mathcal{E}_i(X)B_i + L_i\right)\left(B_i^*\mathcal{E}_i(X)B_i + R_i\right)^{-1}\left(B_i^*\mathcal{E}_i(X)A_i + L_i^*\right). \end{aligned}$$

6.4.2 $\mathcal{H}_\infty$ Control

The $\mathcal{H}_\infty$ control problem consists of finding a controller that stabilizes the linear system (6.9) and ensures that the norm from the additive input disturbance to the output is less than a pre-specified attenuation value γ:

$$\|\mathcal{Z}(\Theta_0,w)\|_\infty = \sup_{w\in l_2^r}\frac{\|\mathcal{Z}(\Theta_0,w)\|_2}{\|w\|_2} < \gamma,$$

where $\mathcal{Z}(\Theta_0,w) = z = (z(0),z(1),z(2),\ldots)$, $x(0) = 0$, $Q_i = C_i^*C_i$, (C,A) detectable in the quadratic mean, $\gamma > 0$ is fixed, $D_i^*D_i = I$, and $C_i^*D_i = 0$. There exists $F = (F_1,\ldots,F_N) \in \mathbb{H}^{n,m}$ which solves this problem, if there exists $X = (X_1,\ldots,X_N) \in \mathbb{H}^{n+}$ satisfying the conditions:

1. $I - \frac{1}{\gamma^2}W_i^*\mathcal{E}_i(X)W_i > 0$;
2. $X_i = Q_i + A_i^*\mathcal{E}_i(X)A_i - A_i^*\mathcal{E}_i(X)\begin{bmatrix} B_i & \frac{1}{\gamma}W_i\end{bmatrix}\bar{R}_i^{-1}\begin{bmatrix} B_i^* \\ \frac{1}{\gamma}W_i^*\end{bmatrix}\mathcal{E}_i(X)A_i$, where

$$\bar{R}_i = \left(\begin{bmatrix} I & 0 \\ 0 & -I \end{bmatrix} + \begin{bmatrix} B_i^* \\ \frac{1}{\gamma} W_i^* \end{bmatrix} \mathcal{E}_i(X) \begin{bmatrix} B_i & \frac{1}{\gamma} W_i \end{bmatrix} \right);$$

3. $r_\sigma(\mathcal{L}) < 1$, where $\mathcal{L}(.) = (\mathcal{L}_1(.), \ldots, \mathcal{L}_N(.))$ is defined by

$$\mathcal{L}_i(.) = \left(A_i + B_i F_i + \frac{1}{\gamma} W_i G_i \right)^* \mathcal{E}_i(.)(\bullet),$$

where

$$G_i = \left(I - \frac{1}{\gamma^2} W_i^* \mathcal{E}_i(X) W_i \right)^{-1} \frac{1}{\gamma^2} W_i^* \mathcal{E}_i(X)(A_i + B_i F_i).$$

$\mathcal{H}_\infty$ control law: The control law is given by

$$u(k) = F_i x(k),$$

with F_i given by

$$F_i = -\left(I + B_i^* \mathcal{E}_i(X) B_i + \frac{1}{\gamma^2} B_i^* \mathcal{E}_i(X) W_i \left(I - \frac{1}{\gamma^2} W_i^* \mathcal{E}_i(X) W_i \right)^{-1} W_i^* \mathcal{E}_i(X) B_i \right)^{-1}$$

$$\times B_i^* \left(I - \frac{1}{\gamma^2} \mathcal{E}_i(X) W_i \left(I - \frac{1}{\gamma^2} W_i^* \mathcal{E}_i(X) W_i \right)^{-1} W_i^* \right) \mathcal{E}_i(X) A_i.$$

6.4.3 Mixed $\mathcal{H}_2/\mathcal{H}_\infty$ Control

Given $\gamma > 0$, the mixed $\mathcal{H}_2/\mathcal{H}_\infty$ control problem is to find a gain $F = (F_1, \ldots, F_N)$ such that when $u(k) = F_{\Theta(k)} x(k)$ the system (6.9) is robustly stable and ζ is minimized, subject to $\|\mathcal{Z}(\Theta_0, w)\|_2 \le \zeta$ and $\|\mathcal{Z}(\Theta_0, w)\|_\infty \le \gamma$. We assume that the transition probability matrix P is not exactly known, but belongs to an appropriate convex set:

$$\mathbb{P} = \left\{ P;\ P = \sum_{k=1}^{M} \alpha_k P^k, \text{ with } \alpha_k \ge 0, \sum_{k=1}^{M} \alpha_k = 1 \right\},$$

where $P^k = [p_{ij}^k]$, $k = 1, \ldots, M$, are known transition probability matrices. The proposed approximation can be handled with the following convex problem subject to linear matrix inequalities. Set

$$\Gamma_i^k = \left[\sqrt{p_{i1}^k} I \cdots \sqrt{p_{iN}^k} I \right] \in \mathbb{B}(\mathbb{C}^{Nn}, \mathbb{C}^n),$$

for $i = 1, \ldots, N$ and $k = 1, \ldots, M$. Given γ^2, find $X = (X_1, \ldots, X_N) \in \mathbb{G}^n$, $Q = (Q_1, \ldots, Q_N) \in \mathbb{G}^n$, $L = (L_1, \ldots, L_N) \in \mathbb{G}^n$, and $Y = (Y_1, \ldots, Y_N) \in \mathbb{G}^{n,m}$, where $\mathbb{G}^{m,n}$ denotes the linear space made up of all sequences of real matrices $U = (U_1, \ldots, U_N)$, with $U_i \in \mathbb{B}(\mathbb{R}^m, \mathbb{R}^n)$ for $i = 1, \ldots, N$. The following Markovian controller combines the performance of the $\mathcal{H}_2$ criterion with the robustness of the $\mathcal{H}_\infty$ one:

Mixed $\mathcal{H}_2/\mathcal{H}_\infty$ control law:

$$\zeta = \min \operatorname{tr}\left\{\sum_{i=1}^{N} W^* X_i W\right\}$$

subject to

$$\begin{bmatrix} Q_i & a_{12} & Q_i C_i^* & Y_i^* D_i^* & W \\ a_{21} & L_i & 0 & 0 & 0 \\ C_i Q_i & 0 & I & 0 & 0 \\ D_i Y_i & 0 & 0 & I & 0 \\ W^* & 0 & 0 & 0 & \gamma^2 I \end{bmatrix} \geq 0, \quad \begin{bmatrix} L_i & L_i \Gamma_i^t \\ \Gamma_i^{t*} L_i & \operatorname{diag}\{Q\} \end{bmatrix} \geq 0, \quad \begin{bmatrix} X_i & I \\ I & Q_i \end{bmatrix} \geq 0,$$

$$X_i = X_i^* > 0, Q_i = Q_i^* > 0, \text{and } L_i = L_i^* > 0,$$

where $tr(.)$ is the trace of a matrix, $\operatorname{diag}\{Q\} \in \mathbb{B}(\mathbb{R}^{Nn})$ is the matrix formed by $Q_1, \ldots, Q_N$ in the diagonal and zeros elsewhere, $a_{12} = Q_i A_i^* + Y_i^* B_i^*$, and $a_{21} = A_i Q_i + B_i Y_i$. If this minimization problem has a solution X, Q, L and Y, then

$$F_i = Y_i Q_i^{-1}.$$

6.5 MJLS $\mathcal{H}_\infty$ Output Feedback Control

The output feedback $\mathcal{H}_\infty$ control for MJLS presented in this section was originally developed in [8]. Consider a continuous-time homogeneous Markov chain, $\Theta = \{\Theta(t) : t > 0\}$, with transition probability $Pr(\theta_{t+\Delta t} = j | \theta_t = i)$ defined as:

$$Pr(\Theta(t+\Delta t) = j | \Theta(t) = i) = \left\{ \begin{array}{ll} \lambda_{ij}(t)\Delta + o(\delta) & \text{if } i \neq j \\ 1 + \lambda_{ii}(t)\Delta + o(\delta) & \text{if } i = j \end{array} \right\}, \tag{6.11}$$

where $\Delta > 0$, and $\lambda_{ij}(t) \geq 0$ is the transition rate of the Markovian state i to $j (i \neq j)$, and

$$\lambda_{ii}(t) = -\sum_{j=1,j\neq i}^{N} \lambda_{ij}(t).$$

The probability distribution of the Markov chain at the initial time is given by $\mu = (\mu_1, \ldots, \mu_N)$, so that $Pr(\Theta(0) = i) = \mu_i$. The MJLS is given by:

$$\begin{aligned} \dot{x}(t) &= A_{\theta(t)}x(t) + B_{\theta(t)}u(t) + W_{\theta(t)}w(t), \\ z(t) &= C_{1_{\theta(t)}}x(t) + D_{1_{\theta(t)}}u(t), \\ y(t) &= C_{2_{\theta(t)}}x(t) + D_{2_{\theta(t)}}w(t), \quad t \geq 0, \end{aligned} \tag{6.12}$$

where the parameters are collections of real matrices,

$$\begin{aligned} A_{\Theta(t)} &= (A_1, \ldots, A_N), \quad \dim(A_i) = n \times n, \\ W_{\Theta(t)} &= (W_1, \ldots, W_N), \quad \dim(W_i) = n \times m, \\ B_{\Theta(t)} &= (B_1, \ldots, B_N), \quad \dim(B_i) = n \times r, \\ C_{1_{\Theta(t)}} &= (C_{11}, \ldots, C_{1N}), \quad \dim(C_{1i}) = p \times n, \\ D_{1_{\Theta(t)}} &= (D_{11}, \ldots, D_{1N}), \quad \dim(D_{1i}) = p \times r, \\ C_{2_{\Theta(t)}} &= (C_{21}, \ldots, C_{2N}), \quad \dim(C_{2i}) = q \times n, \text{ and} \\ D_{2_{\Theta(t)}} &= (D_{21}, \ldots, D_{2N}), \quad \dim(D_{2i}) = q \times m, \quad i = 1, \ldots, N, \end{aligned}$$

with $w \in \mathcal{L}_2(0,T)$ and $E(|x_0|^2) < \infty$. The vectors $x = \{x(t), t \geq 0\}$, $z = \{z(t), t \geq 0\}$, and $y = \{y(t), t \geq 0\}$, are respectively, the state, the controlled output, and the measured output of (6.12). Thus, whenever $\Theta(t) = i \in S$ (where S is an index set $S = \{1, \ldots, N\}$), one has $A_{\theta(t)} = A_i$, $W_{\theta(t)} = W_i$, $B_{\theta(t)} = B_i$, $C_{1_{\theta(t)}} = C_{1i}$, $D_{1_{\theta(t)}} = D_{1i}$, $C_{2_{\theta(t)}} = C_{2i}$, and $D_{2_{\theta(t)}} = D_{2i}$. The output feedback $\mathcal{H}_\infty$ problem for MJLS we consider in this chapter is to find a dynamic controller such that the $\mathcal{H}_\infty$ norm of the closed-loop system is smaller than γ. To find this controller the following linear matrix inequalities must be solved:

$$\begin{bmatrix} \bar{X}_i & X_iW_i + L_iD_{2i} \\ W_i^TX_i + D_{2i}^TL_i^T & -\gamma^{-2}I \end{bmatrix} < 0,$$

$$\begin{bmatrix} \bar{Y}_i & Y_iC_{1i}^T + F_i^TD_{1i}^T & R_i(Y) \\ C_{1i}Y_i + D_{1i}F_i & -I & 0 \\ R_i^T(Y) & 0 & S_i(Y) \end{bmatrix} < 0,$$

$$\begin{bmatrix} Y_i & I \\ I & X_i \end{bmatrix} > 0,$$

with

$$\bar{X}_i = A_i^T X_i + X_i A_i + L_i C_{2i} + C_{2i}^T L_i^T + C_{1i}^T C_{1i} + \sum_{j=i}^{N} \lambda_{ij} X_j,$$

$$\bar{Y}_i = A_i Y_i + Y_i A_i^T + B_i F_i + F_i^T B_i^T + \lambda_{ii} Y_i + \gamma^{-2} W_i W_i^T,$$

$$R_i(Y) = \left[\sqrt{\lambda_{1i}} Y_i, \ldots, \sqrt{\lambda_{(i-1)i}} Y_i, \sqrt{\lambda_{(i+1)i}} Y_i, \ldots, \sqrt{\lambda_{Ni}} Y_i\right],$$

$$S_i(Y) = -\text{diag}(Y_1, \ldots, Y_{i-1}, Y_{i+1}, \ldots, Y_N).$$

Output feedback $\mathcal{H}_\infty$ control law:

$$\dot{v}(t) = A_{c\theta(t)} v(t) + B_{c\theta(t)} y(t),$$
$$u(t) = C_{c\theta(t)} v(t), \; t \geq 0,$$

where

$$A_c = (A_{c1}, \ldots, A_{cN}), \quad \dim(A_{ci}) = n \times n,$$
$$B_c = (B_{c1}, \ldots, B_{cN}), \quad \dim(B_{ci}) = n \times m, \quad \text{and}$$
$$C_c = (C_{c1}, \ldots, C_{cN}), \quad \dim(C_{ci}) = p \times n.$$

$$C_{ci} = F_i Y_i^{-1},$$
$$B_{ci} = (Y_i^{-1} - X_i)^{-1} L_i,$$
$$A_{ci} = (Y_i^{-1} - X_i)^{-1} M_i Y_i^{-1},$$

$$M_i = -A_i^T - X_i A_i Y_i - X_i B_i F_i - L_i C_{2i} Y_i - C_{1i}^T (C_{1i} Y_i + D_{1i} F_i)$$
$$- \gamma^{-2} (X_i W_i + L_i D_{2i}) W_i^T - \sum_{j=1}^{N} \lambda_{ij} Y_j^{-1} Y_i.$$

As before, this controller is based on a continuous MJLS and must be discretized to be implemented in the actual manipulator.

6.6 Examples

The examples presented in this chapter provide guidelines to implement the Markovian controllers on the UARM. These guidelines include the selection of the linearization points and the generation of the state space matrices of the manipulator's linear model. Some details on the probability matrices which describe possible scenarios of faults are also addressed. The fault tolerant system was designed using the Fault Tolerant Manipulator Control Environment option of CERob (Fig. 1.4).

6.6.1 Design Procedures

The graphical interface of the Fault Tolerant option of the simulator shows the Markov jump model (Fig. 6.4) with the current fault configuration highlighted and the Markov chain. While the simulation takes place, the fault configuration changes according to the Markov state, defined by the fault detection system.

The $\mathcal{H}_2$ Markovian controller is selected as the default option. The user may also select the $\mathcal{H}_\infty$, mixed $\mathcal{H}_2/\mathcal{H}_\infty$ or output feedback $\mathcal{H}_\infty$ Markovian controllers on the *Markovian Control* menu and redesign them by pressing the *Controller Design* button (Fig. 6.5).

The Controller Design box is sub-divided in the following options:

- *Controller Type*: specifies the Markovian controller to be designed. The default is an $\mathcal{H}_2$ Markovian controller.
- *Fault Sequence*: specifies the design parameters related to the type of fault sequence selected.
- *Linearization Points*: the equilibrium points around which the robot dynamic model is linearized. The user specifies the size in degrees for the sectors of linearization for all joint variations.
  - *Sector Size*: the linearization point is selected at the middle of the sector whose size is defined at this point. The available values in the examples in this book are: 10°, 20°, 30°, and 40°.
  - *Joint i Variation*: defines the range of variation for each joint. The user selects the minimum and maximum values among the following pre-specified values: −90°, −60°, −40°, −20°, −10°, 0°, 10°, 20°, 40°, 60°, and 90°.
- *Pre-controller Parameters*: defines pre-controller gains K_P and K_D, Eq. 6.4. For each fault sequence, a set of parameters is available to be selected.
- *Fault Probability*: defines the probability of a fault occurrence according to the selected fault sequence.

  The following value can be changed for the AAA–APA fault sequence:

  - *Fault Probability for Joint 2 (%)*: defines the probability of a fault occurrence in the second joint, which changes the robot configuration from AAA to APA.

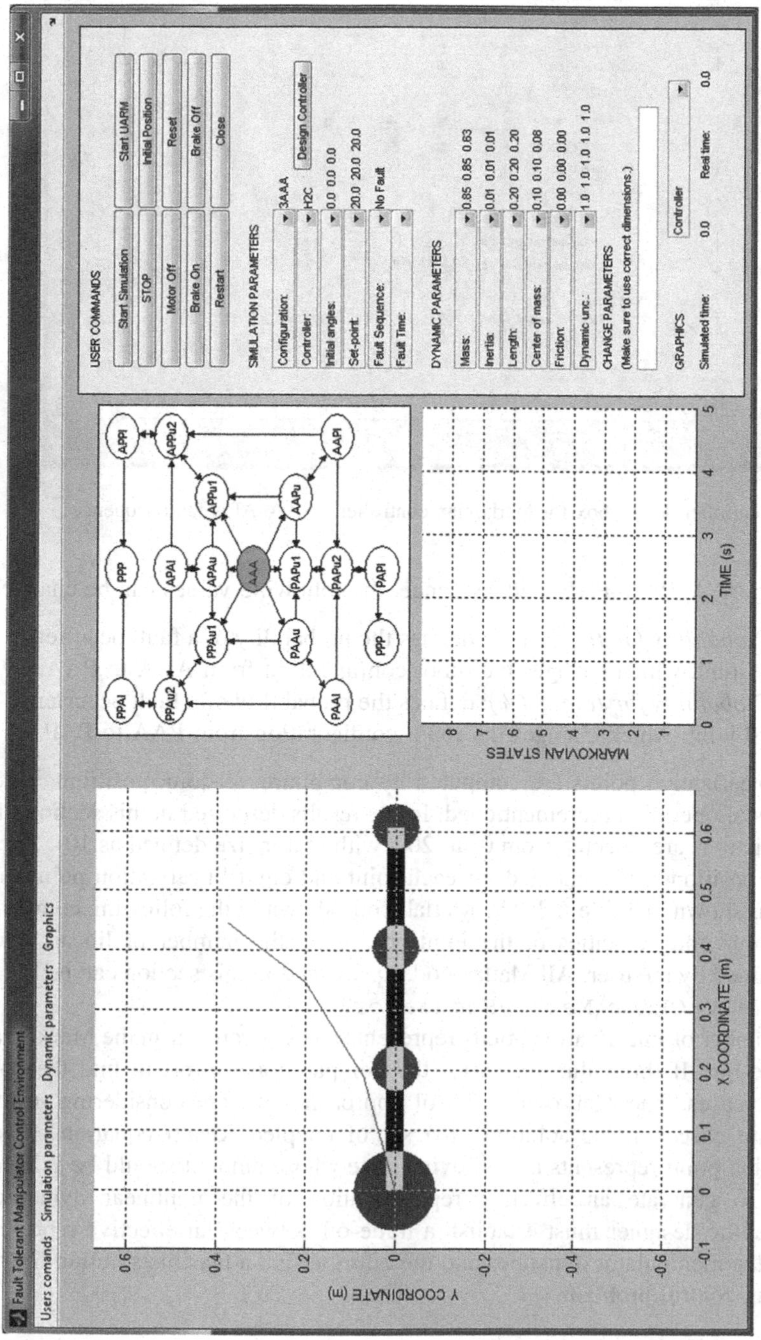

Fig. 6.4 Graphical interface of the Fault Tolerant Manipulator Control Environment

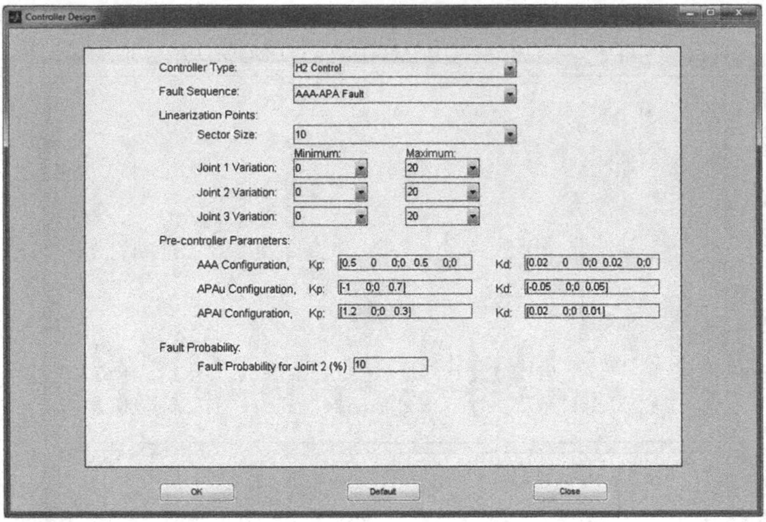

Fig. 6.5 Controller design box for Markovian controllers, AAA–APA fault sequence

For the AAA–PAA–PAP fault sequence, the following values can be changed:

- *Fault Probability for Joint 1 (%)*: defines the probability of a fault occurrence in the first joint, which changes the robot configuration from AAA to PAA.
- *Fault Probability for Joint 3 (%)*: defines the probability of a fault occurrence in the third joint, which changes the robot configuration from PAA to PAP.

The linearization points are computed by combining all joint positions inside the ranges of operation aforementioned. In the results described in this section, the joints variations are selected from $0°$ to $20°$, with sector size defined as $10°$. Then, two joint positions are computed for each joint and eight linearization points are defined, as shown in Table 6.3. The Matlab code shown in the following computes the positions and velocities of the joints based on the number of linearization points defined by the user. All Matlab codes presented in this section can be found in the directory `CERob\MarkovSimulator`.

The number of linearization points represents a key parameter in the Markovian control design. If this value increases, the computational effort to find the controller increases. The Markovian control approaches we are considering in this chapter are based on the solution of a set of coupled Ricatti equations. Each linearization point represents a Markovian state whose amount should be selected properly to generate an effective representation of the nonlinear dynamics. Therefore, the designer must establish a trade-off between an effective representation of the manipulator dynamics and the effort to find a feasible solution for the Markovian control problem.

```
File: Markov_design.m:

   ...

   % compute linearization points
   joint1var = joint1varmax-joint1varmin;
   joint1p = floor(joint1var/sectorsize);
   joint1sectorsize = joint1var/joint1p;
   joint1points =(joint1varmin +...
       joint1sectorsize/2):joint1sectorsize:joint1varmax;

   joint2var = joint2varmax-joint2varmin;
   joint2p = floor(joint2var/sectorsize);
   joint2sectorsize = joint2var/joint2p;
   joint2points =(joint2varmin + ...
       joint2sectorsize/2):joint2sectorsize:joint2varmax;

   joint3var = joint3varmax-joint3varmin;
   joint3p = floor(joint3var/sectorsize);
   joint3sectorsize = joint3var/joint3p;
   joint3points =(joint3varmin + ...
       joint3sectorsize/2):joint3sectorsize:joint3varmax;
   p=0;
   for i = 1:joint1p,
       for j = 1:joint2p,
           for k = 1:joint3p,
               p = p+1;
               q(:,p) = [joint1points(i);...
                         joint2points(j);...
                         joint3points(k)];
           end
       end
   end
   w = zeros(3,p);

   ...
```

For each computed linearization point, the nonlinear dynamics of the robot manipulator is linearized according to Sect. 6.3. The number of Markovian states is related to the number of control phases of a fault sequence, Eq. 6.6. For example, for the AAA–APA fault sequence, the Markovian states related to the three control phases, AAA, APA_u, and APA_l, are computed. The following Matlab code generates the dynamic matrices for each control phase. Although only the computation for the AAA–APA and AAA–PAA–PAP fault sequences are shown, the other possible fault sequences for a three-link manipulator can be reproduced with slight Matlab code modifications.

The system matrices are vectorized in variables *Avec*, *Bvec*, *Wvec*, *Cvec*, *Dvec*, and *Qvec* using the `stack` function, which generates a suitable representation for the MJLS control design. The `stack` function stores a set of matrices in a vector form, where:

- the first element is NaN (Not a Number);
- the second is the number of matrices stored in the vector, N;
- the following $2N$ elements represent the matrices' dimensions;
- the following elements contain the elements of the matrices (columnwise).

The `unstack` function is used to recover the system matrices from vectorized variables.

File: Markov_design.m

```
...

% compute dynamic matrices for each control phase
if fault_seqdesign == 2
    [Avec_1,Bvec_1,Cvec_1,Dvec_1,Wvec_1,Qvec_1] = ...
        AAAd(p,q,w,m,l,lc,I,K,grav,dt);
    [Avec_2,Bvec_2,Cvec_2,Dvec_2,Wvec_2,Qvec_2] = ...
        APA1d(p,q,w,m,l,lc,I,K,grav,dt);
    [Avec_3,Bvec_3,Cvec_3,Dvec_3,Wvec_3,Qvec_3] = ...
        APA2d(p,q,w,m,l,lc,I,K,grav,dt);
    nvec = 3;
elseif fault_seqdesign == 3
    [Avec_1,Bvec_1,Cvec_1,Dvec_1,Wvec_1,Qvec_1] = ...
        AAAd(p,q,w,m,l,lc,I,K,grav,dt);
    [Avec_2,Bvec_2,Cvec_2,Dvec_2,Wvec_2,Qvec_2] = ...
        PAA1d(p,q,w,m,l,lc,I,K,grav,dt);
    [Avec_3,Bvec_3,Cvec_3,Dvec_3,Wvec_3,Qvec_3] = ...
        PAA2d(p,q,w,m,l,lc,I,K,grav,dt);
    [Avec_4,Bvec_4,Cvec_4,Dvec_4,Wvec_4,Qvec_4] = ...
        PAP1d(p,q,w,m,l,lc,I,K,grav,dt);
    [Avec_5,Bvec_5,Cvec_5,Dvec_5,Wvec_5,Qvec_5] = ...
        PAP2d(p,q,w,m,l,lc,I,K,grav,dt);
    [Avec_6,Bvec_6,Cvec_6,Dvec_6,Wvec_6,Qvec_6] = ...
        PAP3d(p,q,w,m,l,lc,I,K,grav,dt);
    nvec = 6;
end
N = nvec*p;

% vectorize dynamic matrices
Astr = []; Bstr = [];
Cstr = []; Dstr = [];
Wstr = []; Qstr = [];
```

```
for i=1:nvec
    istr=int2str(i);
    for j=1:p
        jstr=int2str(j);
        eval(['A',istr,'',jstr,'= '...
            'unstack(Avec_',istr,',',jstr,');']);
        eval(['B',istr,'',jstr,'= '...
            'unstack(Bvec_',istr,',',jstr,');']);
        eval(['C',istr,'',jstr,'= '...
            'unstack(Cvec_',istr,',',jstr,');']);
        eval(['D',istr,'',jstr,'= '...
            'unstack(Dvec_',istr,',',jstr,');']);
        eval(['W',istr,'',jstr,'= '...
            'unstack(Wvec_',istr,',',jstr,');']);
        eval(['Q',istr,'',jstr,'= '...
            'unstack(Qvec_',istr,',',jstr,');']);


        aux=[',A',istr,'',jstr]; Astr=[Astr,aux];
        aux=[',B',istr,'',jstr]; Bstr=[Bstr,aux];
        aux=[',C',istr,'',jstr]; Cstr=[Cstr,aux];
        aux=[',D',istr,'',jstr]; Dstr=[Dstr,aux];
        aux=[',W',istr,'',jstr]; Wstr=[Wstr,aux];
        aux=[',Q',istr,'',jstr]; Qstr=[Qstr,aux];

    end

end;
Astr(1) = [];
eval(['Avec=stack(',Astr,');']);
Bstr(1) = [];
eval(['Bvec=stack(',Bstr,');']);
Cstr(1) = [];
eval(['Cvec=stack(',Cstr,');']);
Dstr(1) = [];
eval(['Dvec=stack(',Dstr,');']);
Wstr(1) = [];
eval(['Wvec=stack(',Wstr,');']);
Qstr(1) = [];
eval(['Qvec=stack(',Qstr,');']);

...
```

The following Matlab code, extracted from the file `AAAd.m`, computes the discretized dynamic matrices for the three-link planar manipulator UARM. Because the UARM moves in a horizontal plane (and therefore is not influenced by gravity) and the linearization points were selected with joint velocities set to zero, the non-inertial term $b(q, \dot{q})$ vanishes. In this case, the dynamic matrix $\widetilde{A}(q, \dot{q})$, defined in Eq. 6.4, is given by:

$$\widetilde{A}(q, \dot{q}) = \left[\begin{matrix} 0 & I \\ \overline{M}^{-1}(q) K_P & \overline{M}^{-1}(q) K_D \end{matrix} \right] \Bigg|_{(q_0, \dot{q}_0)} .$$

Since the Markovian controllers proposed in Sect. 6.4 were developed for discrete time systems, the continuous systems are discretized with a zero-order holder approximation as in the following Matlab code:

File: AAAd.m
```
function [Avec_1,Bvec_1,Cvec_1,Dvec_1,Wvec_1,Qvec_1]=...
    AAAd(p,q,w,m,l,lc,I,K,g,dt)
...
% compute dynamic matrices for each linearized point
for i=1:p,

    aux1 = (q1(i));
    aux2 = (q2(i));
    aux3 = (q3(i));
    aux12 = (q1(i) + q2(i));
    aux23 = (q2(i) + q3(i));
    aux123 = (q1(i) + q2(i)+ q3(i));
    s1 = sin(aux1);
    s2 = sin(aux2);
    c2 = cos(aux2);
    s3 = sin(aux3);
    c3 = cos(aux3);
    s12 = sin(aux12);
    c23 = cos(aux23);
    s23 = sin(aux23);
    c123 = cos(aux123);
    s123 = sin(aux123);

    t1 = lc1*lc1;
    t2 = l1*l1;
    t3 = lc2*lc2;
    t5 = l1*lc2*c2;
    t6 = l2*l2;
    t7 = lc3*lc3;
    t8 = l1*l2*c2;
    t9 = l1*lc3*c23;
    t10 = l2*lc3*c3;

    M(1,1) = I1 + I2 + I3 + m1*t1 + m2*(t2+t3+2.0*t5) + ...
        m3*(t2+t6+t7+2.0*t8+2.0*t9+2.0*t10);
    M(1,2) = I2 + I3 + m2*(t3+t5) + m3*(t6+t7+t8+t9+2.0*t10);
    M(1,3) = I3 + m3*(t7+t9+t10);
    M(2,1) = M(1,2);
    M(2,2) = I2 + I3 + m2*t3 + m3*(t6+t7+2.0*t10);
    M(2,3) = I3 + m3*(t7+t10);
    M(3,1) = M(1,3);
    M(3,2) = M(2,3);
    M(3,3) = I3 + m3*t7;

    W = inv(M);

    % number of joints
    n = 3;

    % write dynamic matrices
    A_temp = [zeros(n) eye(n); W*kp  W*kv];
    B_temp = [zeros(n); W];
```

```
    % define constant matrices
    C1 = [alpha*eye(n) zeros(n);
              zeros(n)   zeros(n)];
    D1 = [zeros(n);beta*eye(n)];

    % discretize dynamic systems
    sistC = ss(A_temp,B_temp,C1,D1);
    sistD = c2d(sistC,d,'zoh')
    [A_temp,B_temp,C1,D1] = ssdata(sistD);
    W1 = B_temp;
    Q1 = M1'*M1;
end
```

The probability matrices described in Eqs. 6.7 and 6.8 can be computed as in the following code for an arbitrary number of linearization points. The probability that the system will stay at a given linearization point is set equally for all points. If the manipulator happens to stay too much time near a specific linearization point, the probability of the system to stay at this point can be increased. The variables `fault1`, `fault2` and `fault3` define, respectively, the fault probability for joints 1, 2, and 3, according to the fault sequence being considered. These values are selected in the Controller Design box, Fig. 6.5.

The Markovian jump controllers are computed using the Discrete Time Markovian Jump Linear Systems (DTMJLS) toolbox [5]. The Matlab codes of the control design functions can be found in the directory `CERob\Markov-Simulator\ dtmjls`.

File: Markov_design.m

```
...

% define probability matrix
P0 = zeros(p);

%AAA-APA fault sequence
if fault_seqdesign == 2

    Pf = fault2*eye(p);
    Ps = 2*Pf;

    Paux = [zeros(p) Pf P0;
        P0 zeros(p) Ps;
        P0 Ps zeros(p)];
    vaux = (ones(N,1)-sum(Paux,2))/(p+2);

    Paux2 = [vaux(1)*ones(p) zeros(p,N-p);
        zeros(p,p) vaux(1+p)*ones(p) zeros(p,N-2*p);
        zeros(p,2*p) vaux(1+2*p)*ones(p) zeros(p,N-3*p)];

%AAA-PAA-PAP fault sequence
elseif fault_seqdesign == 3
    Pf1 = fault1*eye(p);
    Pf3 = fault3*eye(p);
    Ps = 2*Pf1;
```

```
    Paux = [zeros(p) Pf1 P0 Pf3 P0 P0;
        P0 zeros(p) Ps Pf3 P0 P0;
        P0 Ps zeros(p) P0 Pf3 P0;
        P0 P0 P0 zeros(p) Ps P0;
        P0 P0 P0 Ps zeros(p) Ps;
        P0 P0 P0 Ps Ps zeros(p)];

    vaux = (ones(N,1)-sum(Paux,2))/(p+2);
    Paux2 = [vaux(1)*ones(p) zeros(p,N-p);
        zeros(p,p) vaux(1+p)*ones(p) zeros(p,N-2*p);
        zeros(p,2*p) vaux(1+2*p)*ones(p) zeros(p,N-3*p);
        zeros(p,3*p) vaux(1+3*p)*ones(p) zeros(p,N-4*p);
        zeros(p,4*p) vaux(1+4*p)*ones(p) zeros(p,N-5*p);
        zeros(p,5*p) vaux(1+5*p)*ones(p);];

end

P = diag(2*vaux) + Paux +Paux2;

% compute Markovian controller
if  all(contdesign == 'H2C') %H2 controller

    mL=zeros(6,3); mR= 1.5*eye(3);
    mLstr=[];mRstr=[];
    for i=1:N
        aux=[',mL']; mLstr=[mLstr,aux];
        aux=[',mR']; mRstr=[mRstr,aux];
    end;
    mLstr(1)=[];
    eval(['Lvec=stack(',mLstr,');']);
    mRstr(1)=[];
    eval(['Rvec=stack(',mRstr,');']);
    options=[1e-12 1e3 -1 100 1]; v=1;
    [Fvec,Xvec] = ...
        mkvlqr(Avec,Bvec,Qvec,Lvec,Rvec,P,options,v);

elseif all(contdesign == 'HIN') %Hinf controller

    nu = 10;
    opt1=[0 0 0 0 0]; opt2=[1.e-5 4000 1];
    [Fvec,Xvec] = ...
        hinfsopt(Avec,Bvec,Wvec,Cvec,Dvec,P,nu,opt1,opt2);

elseif  all(contdesign == 'MIX') %mixed H2/Hinf controller

    nu = 1000;
    W = unstack(Wvec,1);
    v = 1; options=[1e-2 1e3 -1 10 0];
    Fvec = mixopt(Avec,Bvec,W,Cvec,Dvec,P,nu,options,v);

end
```

6.6.2 AAA–APA Fault Sequence: State-Feedback Control

The following experimental results are obtained by introducing a fault on the second joint of UARM. The results were obtained for the same initial and final positions of Sect. 6.2, $q(0) = [0\ 0\ 0]^\circ$ and $q(T) = [20\ 20\ 20]^\circ$, respectively. The initial configuration of the manipulator is AAA, with the Markovian state starting in 1 (Table 6.3). The changes of the Markov states with respect to the linearization points are defined according to the actual position of the manipulator's joints. The joint position ranges and sector sizes, selected in the Control Design box, are considered in the Markov state computation.

The fault is introduced at $t_f = 2.5$ s. At the detection time t_d, the Markovian chain changes from the configuration AAA to the control phase APA_u. The Markovian chain changes from the control phase APA_u to the control phase APA_l when the second joint reaches the set-point at time t_r (Fig. 6.7).

Torque disturbances of the following form are introduced in each joint to test the robustness of the controllers:

$$\tau_d = \begin{bmatrix} 0.03e^{-2(t-t_f)^2}\sin(4\pi t) \\ 0.015e^{-2(t-t_f)^2}\sin(5\pi t) \\ 0.009e^{-2(t-t_f)^2}\sin(6\pi t) \end{bmatrix}.$$

These disturbances are sinusoidal oscillations attenuated by normal functions (Fig. 6.6). The disturbance in the passive joint is turned off after the fault occurs.

The preliminary PD controllers are used jointly with the Markovian controllers to pre-compensate the model imprecisions. For the specific case of planar manipulators, these controllers are useful to generate a more representative state space model of the system. From Sect. 6.6.1, it is clear that the dynamic matrix A

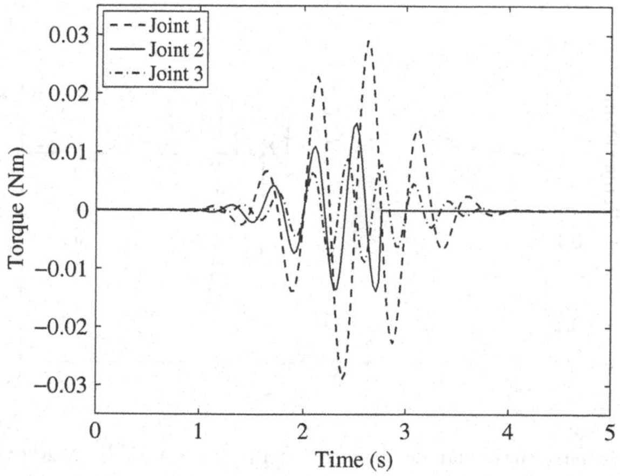

Fig. 6.6 Torque disturbances

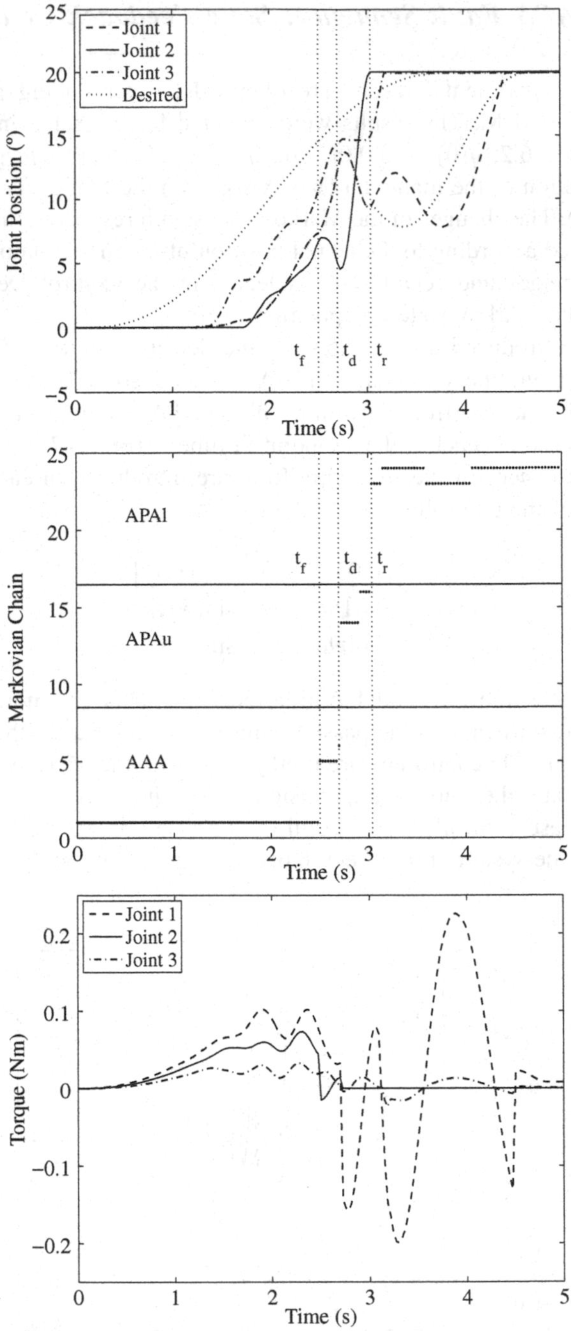

Fig. 6.7 Joint positions, Markovian chain states, and joint torques, $\mathcal{H}_2$ Markovian control (t_f = fault time, t_d = detection time, and t_r = time when the second joint reaches the desired position)

will be the same for all linearization points if the PD gains are set to zero. These gains are selected individually for each robot configuration, in order to obtain the best performance for a complete desired trajectory. For the state-feedback Markovian controllers in this section, the PD gains were selected heuristically as:

$$K_{P_{\mathrm{AAA}}} = \begin{bmatrix} 0.2 & 0 & 0 \\ 0 & 0.15 & 0 \\ 0 & 0 & 0.12 \end{bmatrix}, \; K_{D_{\mathrm{AAA}}} = \begin{bmatrix} 0.02 & 0 & 0 \\ 0 & 0.02 & 0 \\ 0 & 0 & 0.02 \end{bmatrix},$$

$$K_{P_{\mathrm{APA}_l}} = \begin{bmatrix} 1.25 & 0.20 \\ 0.06 & 0.30 \end{bmatrix}, \; K_{D_{\mathrm{APA}_l}} = \begin{bmatrix} 0.27 & 0.02 \\ 0.01 & 0.01 \end{bmatrix},$$

$$K_{P_{\mathrm{APA}_u}} = \begin{bmatrix} -1.10 & -0.05 \\ -0.07 & 0.7 \end{bmatrix}, \; K_{D_{\mathrm{APA}_u}} = \begin{bmatrix} -0.07 & -0.01 \\ -0.04 & 0.06 \end{bmatrix}.$$

The Markovian controllers are computed considering $\alpha = 20$ for configuration AAA; $\alpha = 40$ for the control phases APA_u and APA_l; and $\beta = 1$ for all configurations (Eq. 6.4). Note that the conditions $\widetilde{C}^T\widetilde{D} = 0$ and $\widetilde{D}^T\widetilde{D} = I$ are satisfied. For the $\mathcal{H}_2$ Markovian control, the weighting matrices are defined by $Q = \widetilde{C}^T\widetilde{C}$ and $R = I$. The best value of γ for the $\mathcal{H}_\infty$ and mixed $\mathcal{H}_2/\mathcal{H}_\infty$ Markovian controllers is $\gamma = 10$.

Fault detection and isolation is performed by the filtered torque prediction error approach proposed in [9]. In essence, this procedure is based on the prediction error signal, $\epsilon(t)$, between the filtered torque given by

$$\dot{\tau}_f = -\lambda\tau_f + \eta\tau, \quad \tau_f(0) = 0,$$

where η and λ are positive filter constants, and the filtered torque estimate given by

$$\hat{\tau}_f = Y_f\hat{\theta},$$

where Y_f is the filtered regression matrix and $\hat{\theta}$ is a constant, best-guess estimate for θ, the vector containing the uncertain system parameters. From experimental verification, the fault detection parameters that provide the lowest delay times between fault occurrence and fault detection, without indicating false detections, are $\eta = 1$, $\lambda = 10$ and $\gamma_2 = [0.007 \;\; 0.005 \;\; 0.004]$ (see [9] for more details about the parameter γ_2). The mean delay time between the fault occurrence and the detection, for all controllers designed in this section, is 213 ms.

The experimental results, including joint positions, Markovian chains, and torques for $\mathcal{H}_2$, $\mathcal{H}_\infty$, and mixed $\mathcal{H}_2/\mathcal{H}_\infty$ Markovian controllers are shown in Figs. 6.7, 6.8 and 6.9.

The effectiveness of the fault tolerant system presented in this chapter can be verified in the results displayed in Fig. 6.10, which shows an experiment where the fault in the second joint is introduced at $t_f = 2.3$ s and detected at $t_f = 2.9$ s, with a delay of 600 ms.

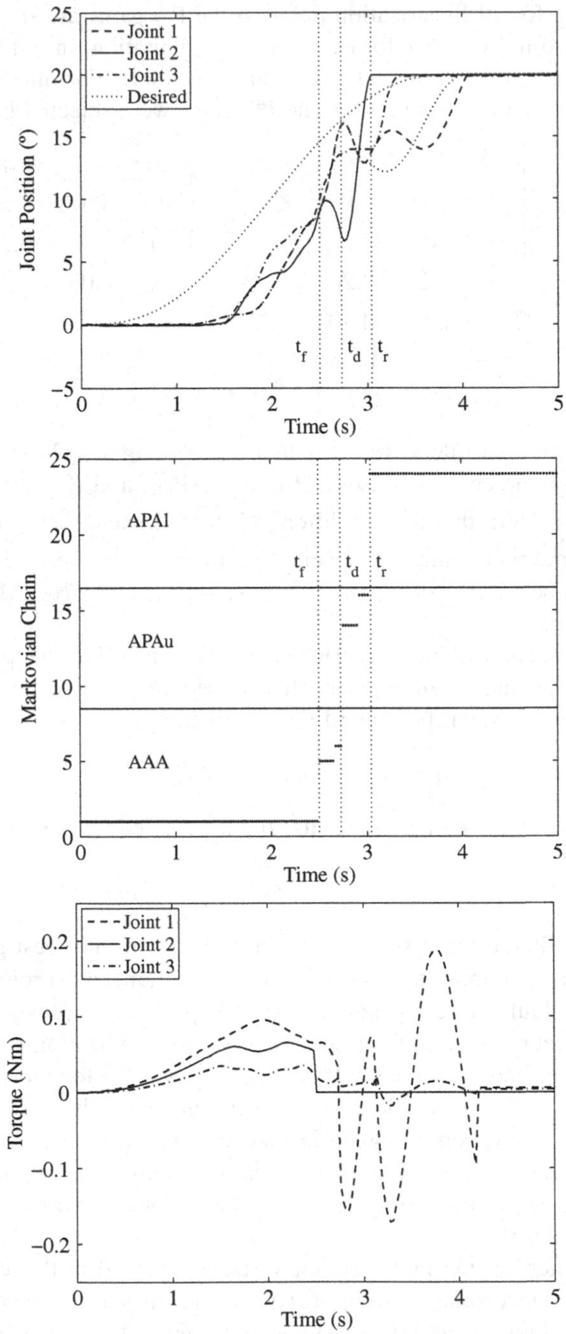

Fig. 6.8 Joint positions, Markovian chain states, and joint torques, $\mathcal{H}_\infty$ Markovian control

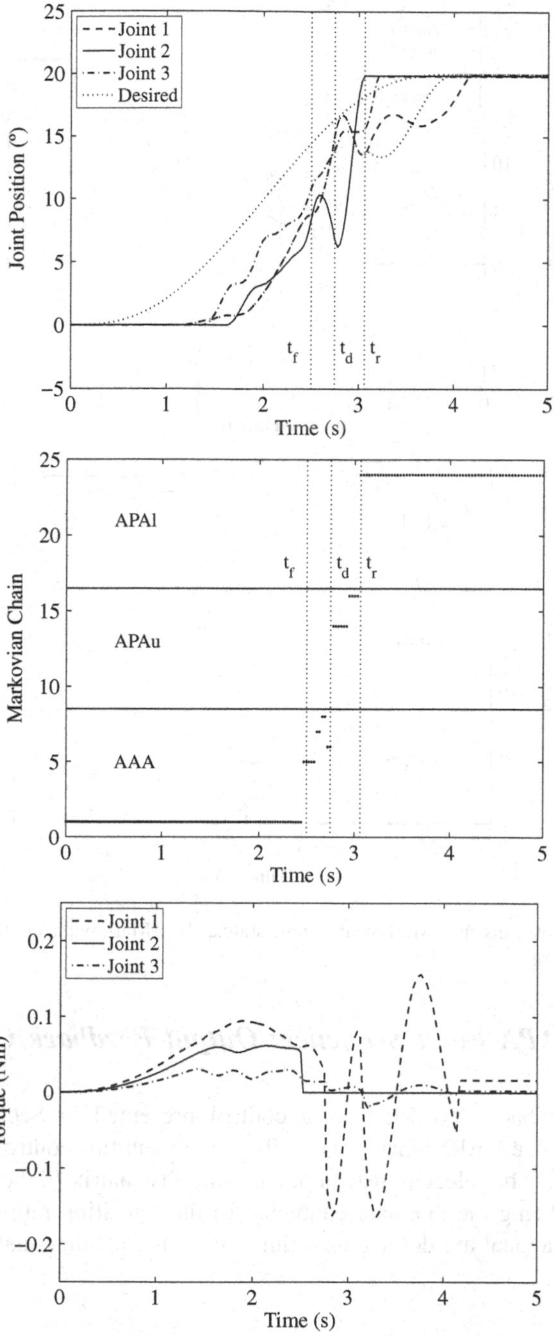

Fig. 6.9 Joint positions, Markovian chain states, and joint torques, mixed $\mathcal{H}_2/\mathcal{H}_\infty$ Markovian control

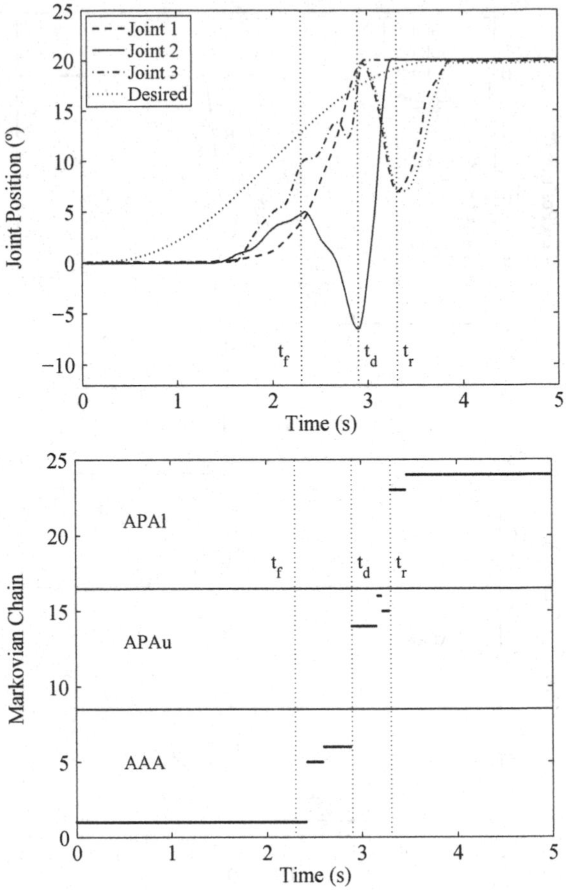

Fig. 6.10 Joint positions and Markovian chain states, $\mathcal{H}_\infty$ Markovian control, delay time to detect fault = 600 ms

6.6.3 AAA–APA Fault Sequence: Output-Feedback Control

The output feedback $\mathcal{H}_\infty$ Markovian control presented in Sect. 6.5 was also implemented in the UARM manipulator. To implement this controller, a transition rate matrix Λ may be selected instead of a probability matrix P, the only difference between them being the diagonal elements. In the transition rate matrix, the elements of the diagonal are defined as a function of the off-diagonal elements as in (6.11), that is,

$$\lambda_{ii}(t) = -\sum_{j=1, j\neq i}^{N} \lambda_{ij}(t).$$

For the AAA–APA fault sequence, the 24×24 transition rate matrix Λ is partitioned into 9 submatrices of dimension 8×8, similarly to what was done for matrix P in Sect. 6.3.4:

$$\Lambda = \left[\begin{array}{c|c|c} \Lambda_{AAA} & \Lambda_f & \Lambda_0 \\ \hline \Lambda_0 & \Lambda_{\mathrm{APA}_u} & \Lambda_s \\ \hline \Lambda_0 & \Lambda_s & \Lambda_{\mathrm{APA}_l} \end{array}\right].$$

The submatrix Λ_{AAA} shows the relationship between the linearization points of configuration AAA, and the diagonal submatrix Λ_f determines the probabilities of fault occurrence. In the second line of Λ, Λ_{APA_u} defines the relationship between the linearization points in the control phase APA_u, Λ_0 represents the fact that the free joint cannot be repaired, and the matrix Λ_s represents the transition rate of the system to go to the control phase APA_l. In the third line of Λ, Λ_{APA_l} defines the relationship between the linearization points in the set APA_l, Λ_s represents the probability that the system will return to the control phase APA_u, and Λ_0 represents, again, the impossibility of the free joint being repaired. The transition rate matrix Λ is selected as:

$$\begin{aligned} &\Lambda_{AAA_{(ij)}} = 0.09, \quad \Lambda_{\mathrm{APA}_{u(ij)}} = 0.08, \quad \Lambda_{\mathrm{APA}_{l(ij)}} = 0.08, \quad \text{for } i \neq j, \\ &\Lambda_{AAA_{(ii)}} = -0.73, \Lambda_{\mathrm{APA}_{u(ii)}} = -0.76, \Lambda_{\mathrm{APA}_{l(ii)}} = -0.76, \quad \text{for } i = j, \\ &\Lambda_f = 0.1I_8, \quad \Lambda_s = 0.2I_8, \quad \Lambda_0 = 0. \end{aligned}$$

The experiments considering the output feedback $\mathcal{H}_\infty$ Markovian controller are performed for the same initial and final positions as before, with the fault introduced at $t_f = 1.5$s. Since joint velocities are not available to the preliminary controller, a proportional-only controller is used. The P gains are selected as:

$$K_{P_{AAA}} = \begin{bmatrix} 2.25 & 0 & 0 \\ 0 & 2.0 & 0 \\ 0 & 0 & 1.8 \end{bmatrix}, \quad K_{P_{\mathrm{APA}_u}} = \begin{bmatrix} -1.8 & 0 \\ 0 & 0.5 \end{bmatrix}, \quad K_{P_{\mathrm{APA}_l}} = \begin{bmatrix} 2.0 & 0.3 \\ 0.1 & 0.7 \end{bmatrix}.$$

The controller is computed considering $\alpha = 50$ and $\beta = 100$ for all configurations and the best value of γ is 1.5. The experimental results, including joint positions, Markovian states and joint torques for the output-feedback $\mathcal{H}_\infty$ Markovian controller are shown in Fig. 6.11.

In order to compare the performances of all Markovian controllers, $N = 10$ experiments for each controller are performed to compute the performance indexes $\mathcal{L}_2[x]$ and $E[\tau]$. The results displayed in Figs. 6.7, 6.8 and 6.9 correspond to the samples that are closer to the mean values $\mathcal{L}_2[x]$ and $E[\tau]$. The values of $\mathcal{L}_2[x]$ and $E[\tau]$ for the AAA–APA fault sequence are shown in Table 6.5. The $\mathcal{H}_2$ controller spends more energy with a worst performance than the $\mathcal{H}_\infty$ and the $\mathcal{H}_2/\mathcal{H}_\infty$ controllers. The output-feedback $\mathcal{H}_\infty$ controller spends the largest amount of energy, but also presents the best performance of all controllers.

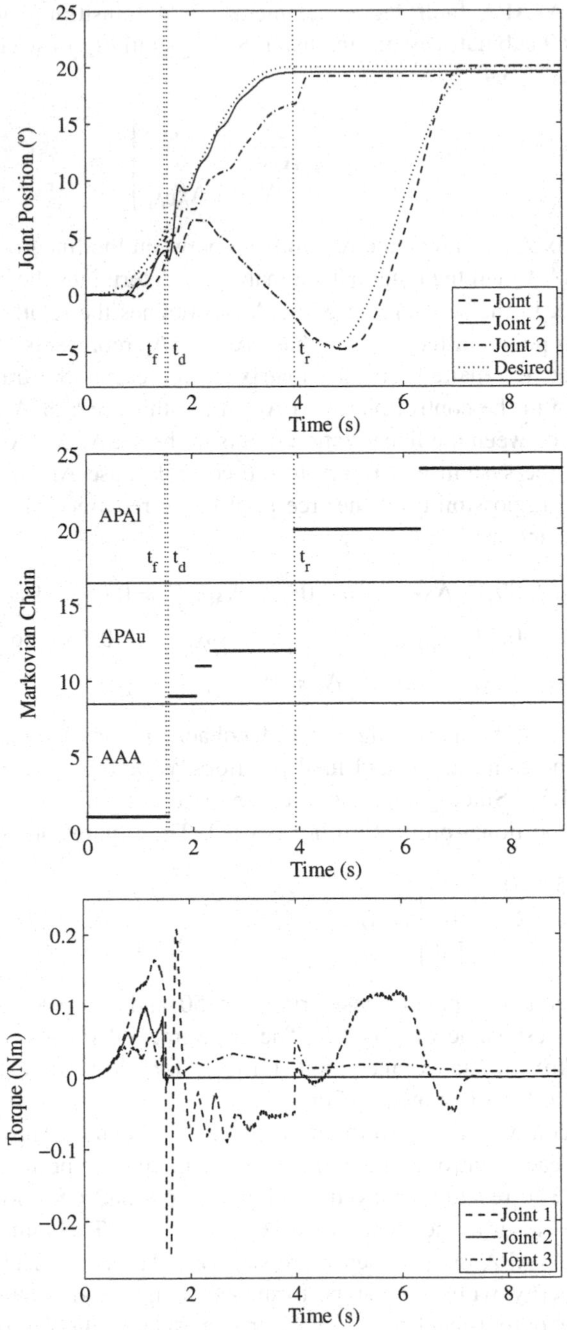

Fig. 6.11 Joint positions, Markovian chain states, and joint torques, output-feedback $\mathcal{H}_\infty$ Markovian control

Table 6.5 Performance indexes—AAA–APA fault sequence

Controller	$\mathcal{L}_2[x]$	$E[\tau]$ (N m s)
$\mathcal{H}_2$ Markovian	0.2243	0.3327
$\mathcal{H}_\infty$ Markovian	0.2093	0.3040
Mixed $\mathcal{H}_2/\mathcal{H}_\infty$ Markovian	0.2071	0.2934
Output-feedback $\mathcal{H}_\infty$ Markovian	0.1316	0.3860

6.6.4 AAA–PAA–PAP Fault Sequence: State-Feedback Control

We present now an experiment where the manipulator is subject to two consecutive faults, AAA–PAA–PAP. Our objective is to motivate the reader to generalize the controller design guidelines to n-link manipulators subject to m faults. Two artificial faults are introduced at $t_{f1} = 2.5$ s and $t_{f2} = 2.7$ s in joints 1 and 3, respectively. At fault detection times t_{d1} and t_{d2}, the Markovian state changes from configuration AAA to control phase PAA$_u$, and from this one to the control phase PAP$_{u1}$. When the first and third joints reach their set-points at times t_{r1} and t_{r2}, the Markovian states changes to the control phases PAA$_{u1}$ and PAP$_l$, respectively.

The experiment is performed considering the same initial and final positions and torque disturbances used in Sect. 6.6.2. For the state-feedback Markovian controllers the PD gains are selected as:

$$K_{P_{\text{AAA}}} = \begin{bmatrix} 0.2 & 0 & 0 \\ 0 & 0.15 & 0 \\ 0 & 0 & 0.12 \end{bmatrix}, \quad K_{D_{\text{AAA}}} = \begin{bmatrix} 0.02 & 0 & 0 \\ 0 & 0.02 & 0 \\ 0 & 0 & 0.02 \end{bmatrix},$$

$$K_{P_{\text{PAA}_l}} = \begin{bmatrix} 1 & 0 \\ 0 & 1 \end{bmatrix}, \quad K_{D_{\text{PAA}_l}} = \begin{bmatrix} 0.01 & 0 \\ 0 & 0.01 \end{bmatrix},$$

$$K_{P_{\text{PAA}_u}} = \begin{bmatrix} -0.5 & 0 \\ 0 & 0.2 \end{bmatrix}, \quad K_{D_{\text{PAA}_u}} = \begin{bmatrix} -0.01 & 0 \\ 0 & 0.05 \end{bmatrix},$$

$$K_{P_{\text{PAP}_{u1}}} = -0.5, \quad K_{D_{\text{PAP}_{u1}}} = -0.01,$$

$$K_{P_{\text{PAP}_{u2}}} = -10, \quad K_{D_{\text{PAP}_{u2}}} = -0.7,$$

$$K_{P_{\text{PAP}_l}} = 2, \quad K_{D_{\text{PAP}_l}} = 0.5.$$

The Markovian controllers are computed considering $\alpha = 20$ for configuration AAA; $\alpha = 40$ for the control phases PAA$_u$ and PAA$_l$; $\alpha = 10$ for control phases PAP$_{u1}$, PAP$_{u2}$ and PAP$_l$; and $\beta = 1$ for all configurations (see Eq. 6.4). These parameters are selected empirically for both fault sequences. The best value of γ for the $\mathcal{H}_\infty$ and mixed $\mathcal{H}_2/\mathcal{H}_\infty$ Markovian controllers is $\gamma = 10$.

The best fault detection parameters for this fault sequence are $\eta = 1$, $\lambda = 10$, and $\gamma_2 = [0.0125 \ 0.0090 \ 0.0075]$. The mean delay time between the first and second fault occurrence and the detection for all controllers of this section are, respectively, 51 ms and 210 ms. The experimental results, including joint positions and Markov chains, are shown in Figs. 6.12, 6.13 and 6.14. Figure 6.15 presents the joint torques for all three controllers.

6.6.5 AAA–PAA–PAP Fault Sequence: Output-Feedback Control

To implement the output-feedback $\mathcal{H}_\infty$ Markovian control, for the AAA–PAA–PAP fault sequence, the transition rate matrix Λ is partitioned in 36 submatrices of dimension 8×8:

$$\Lambda = \left[\begin{array}{c|c|c|c|c|c} \Lambda_{\mathrm{AAA}} & \Lambda_f & \Lambda_0 & \Lambda_f & \Lambda_0 & \Lambda_0 \\ \hline \Lambda_0 & \Lambda_{\mathrm{PAA}_u} & \Lambda_s & \Lambda_f & \Lambda_0 & \Lambda_0 \\ \hline \Lambda_0 & \Lambda_s & \Lambda_{\mathrm{PAA}_l} & \Lambda_0 & \Lambda_f & \Lambda_0 \\ \hline \Lambda_0 & \Lambda_0 & \Lambda_0 & \Lambda_{\mathrm{PAP}_{u1}} & \Lambda_s & \Lambda_0 \\ \hline \Lambda_0 & \Lambda_0 & \Lambda_0 & \Lambda_s & \Lambda_{\mathrm{PAP}_{u2}} & \Lambda_s \\ \hline \Lambda_0 & \Lambda_0 & \Lambda_0 & \Lambda_s & \Lambda_s & \Lambda_{\mathrm{PAP}_l} \end{array}\right].$$

Following the same arguments presented in Sect. 6.6.2, Λ is defined as:

$$\begin{aligned}
&\Lambda_{\mathrm{AAA}(ij)} = 0.08, \Lambda_{\mathrm{PAA}_{u}(ij)} = 0.07, \Lambda_{\mathrm{PAA}_{l}(ij)} = 0.07, \\
&\Lambda_{\mathrm{PAP}_{u1}(ij)} = 0.08, \Lambda_{\mathrm{PAP}_{u2}(ij)} = 0.06, \Lambda_{\mathrm{PAP}_{l}(ij)} = 0.06, \quad \text{for} \quad i \neq j \\
&\Lambda_{\mathrm{AAA}(ii)} = -0.76, \Lambda_{\mathrm{PAA}_{u}(ii)} = -0.79, \Lambda_{\mathrm{PAA}_{l}(ii)} = -0.79, \\
&\Lambda_{\mathrm{PAP}_{u1}(ii)} = -0.76, \Lambda_{\mathrm{PAP}_{u2}(ii)} = -0.82, \Lambda_{\mathrm{PAP}_{l}(ii)} = -0.82, \quad \text{for} \quad i = j, \\
&\Lambda_f = 0.1 I_8, \Lambda_s = 0.2 I_8, \Lambda_0 = 0.
\end{aligned}$$

The fault detection system used in AAA–APA fault sequence is also adopted here to determine the fault occurrence. The first fault is introduced at $t_{f1} = 2.5$ s and detected at $t_{d1} = 2.58$ s, changing the Markovian state from configuration AAA to control phase PAA_u, maintaining its linearization point. The second fault is introduced at $t_{f2} = 3.0$ s and detected at $t_{d2} = 3.35$ s, before joint 1 has reached its set point, forcing the Markovian state to jump from control phase PAA_u to control phase PAP_{u1}. After joint 1 reaches its set point at $t_{r1} = 4.73$ s, the Markovian state changes to control phase PAP_{u2}, and finally at $t_{r3} = 6.75$ s, the state changes to control phase PAP_l. Torque disturbances and an additional payload of 0.5 kg are introduced in order to test controller robustness. The proportional gains are selected as:

$$K_{P_{\mathrm{AAA}}} = \begin{bmatrix} 2.25 & 0 & 0 \\ 0 & 2.0 & 0 \\ 0 & 0 & 1.8 \end{bmatrix}, \quad K_{P_{\mathrm{PAA}_u}} = \begin{bmatrix} -1.8 & 0 \\ 0 & 0.5 \end{bmatrix}, \quad K_{P_{\mathrm{PAA}_l}} = \begin{bmatrix} -1.8 & 0 \\ 0 & 0.5 \end{bmatrix},$$

$$K_{P_{\mathrm{PAP}_{u1}}} = -5, \quad K_{P_{\mathrm{PAP}_{u2}}} = -30, \quad K_{P_{\mathrm{PAP}_l}} = 20.$$

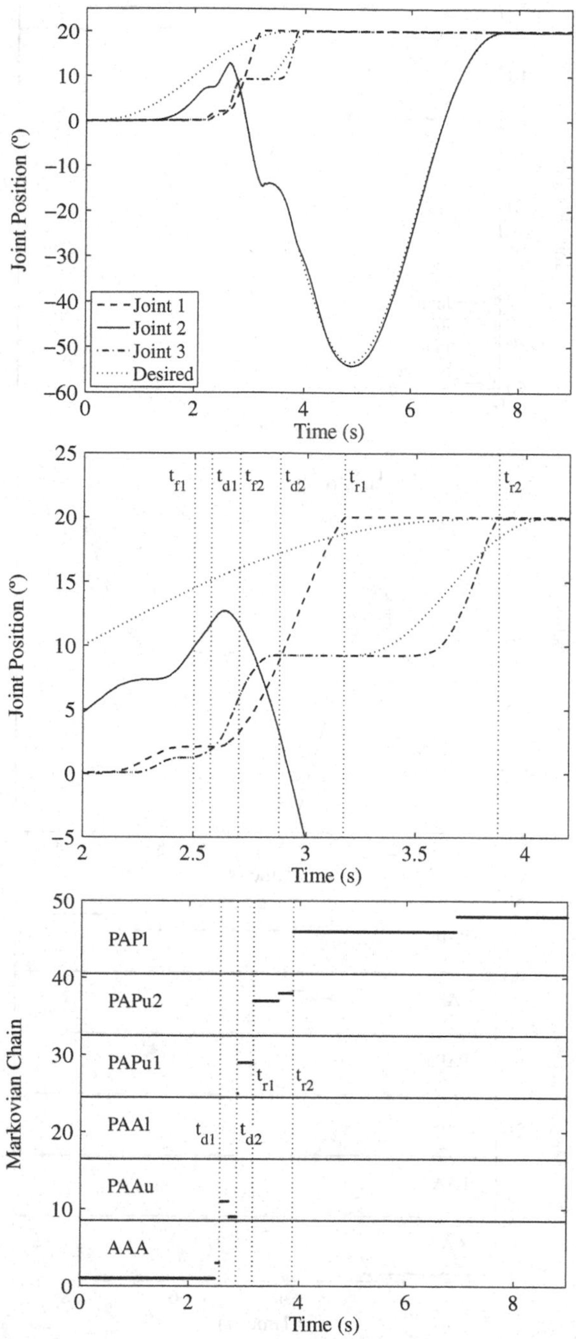

Fig. 6.12 Joint positions and Markovian chain states, $\mathcal{H}_2$ Markovian control. The middle graph is a zoomed-in version of the top graph (t_{f1} = occurrence time of the first fault, t_{d1} = detection time of the first fault, t_{f2} = occurrence time of the second fault, t_{d2} = detection time of the second fault, t_{r1} = time when the first joint reaches the desired position, and t_{r2} = time when the third joint reaches the desired position)

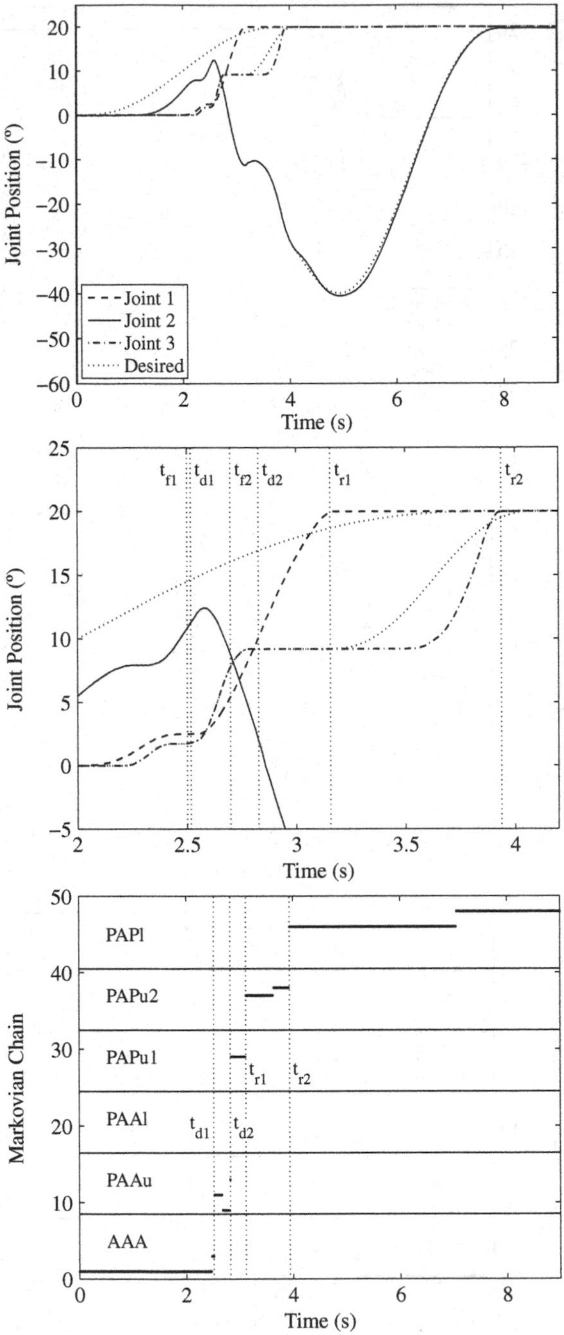

Fig. 6.13 Joint positions and Markovian chain states, $\mathcal{H}_\infty$ Markovian control. The middle graph is a zoomed-in version of the top graph

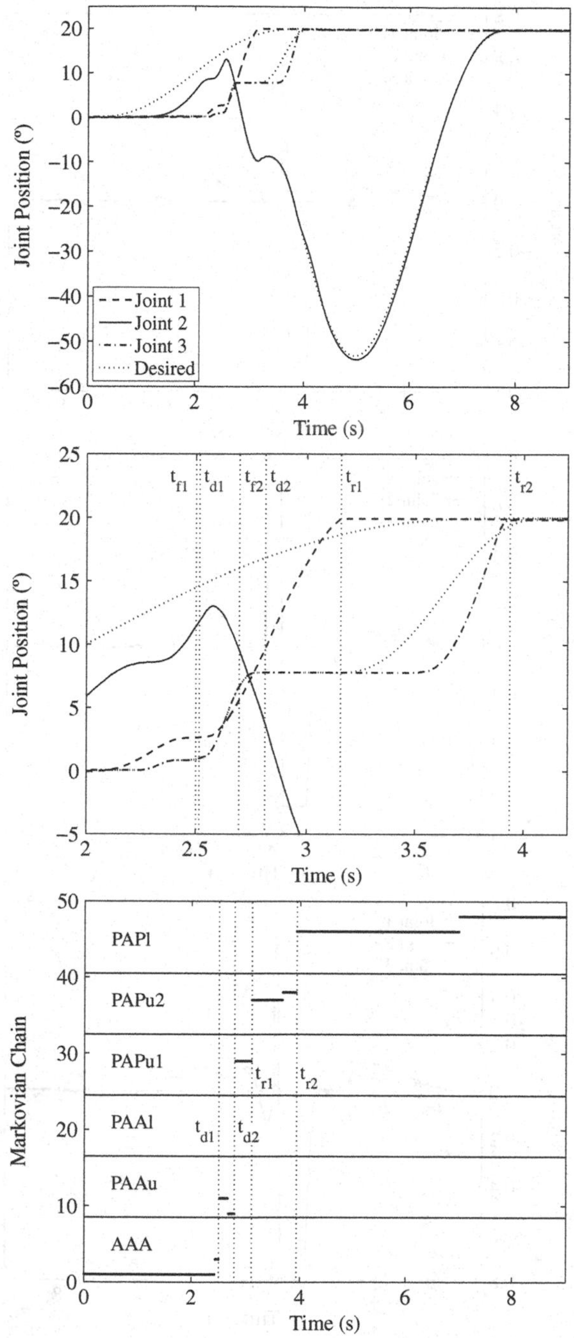

Fig. 6.14 Joint positions and Markovian chain states, mixed $\mathcal{H}_2/\mathcal{H}_\infty$ Markovian control. The middle graph is a zoomed-in version of the top graph

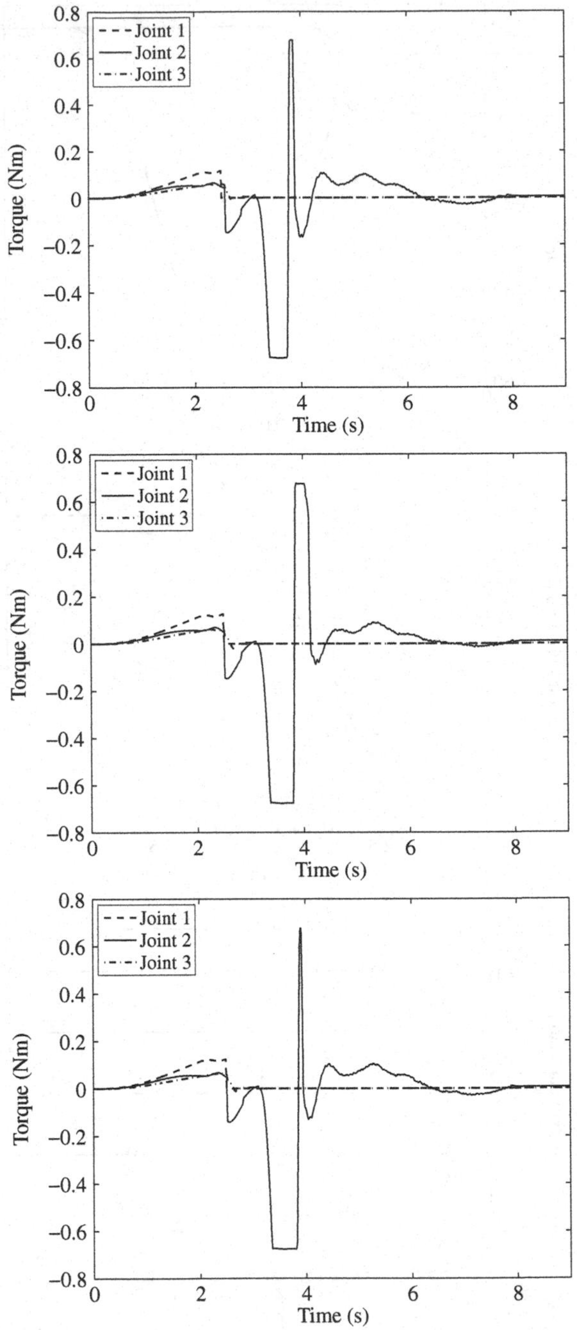

Fig. 6.15 Joint torques, $\mathcal{H}_2$ (*top*), $\mathcal{H}_\infty$ (*middle*), and mixed $\mathcal{H}_2/\mathcal{H}_\infty$ (*bottom*) Markovian controllers

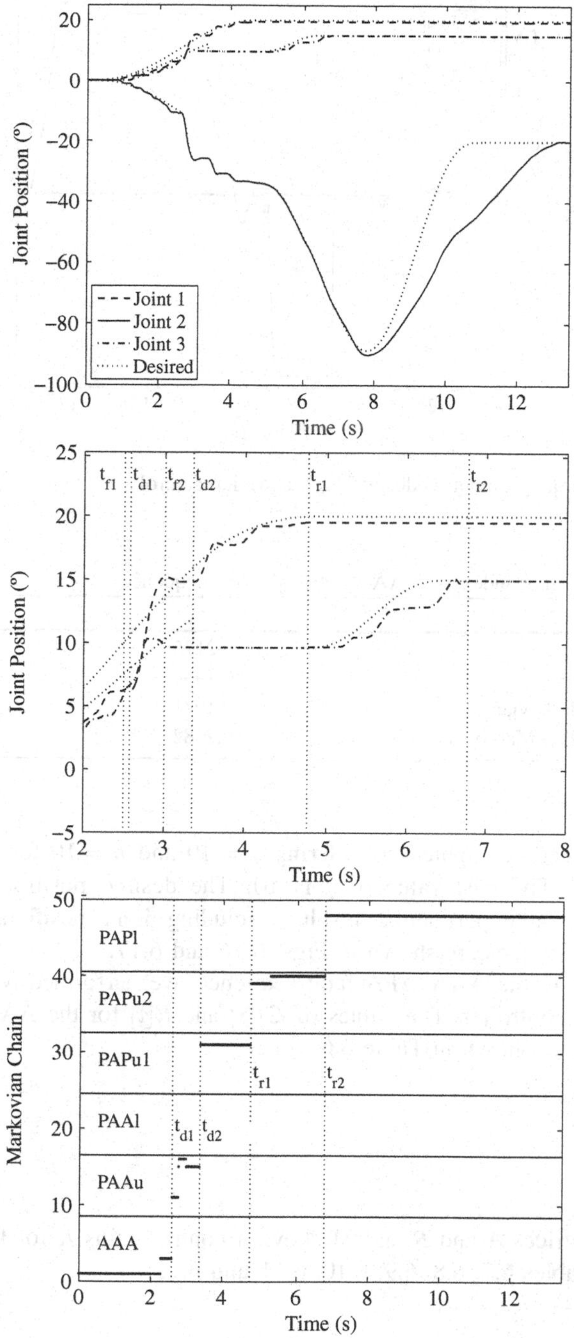

Fig. 6.16 Joint positions and Markovian chain states, output-feedback $\mathcal{H}_\infty$ Markovian control. The middle graph is a zoomed-in version of the top graph

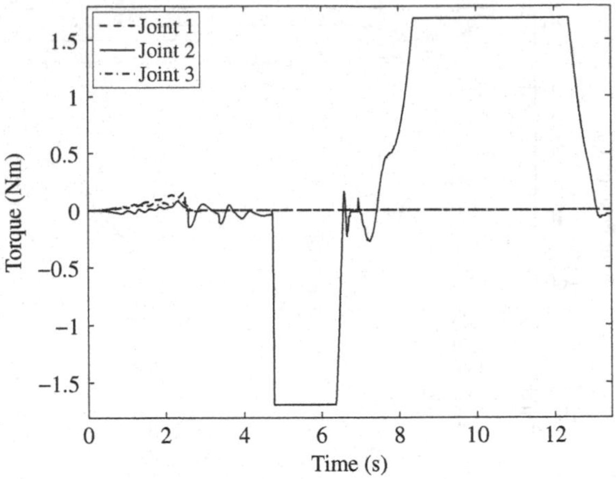

Fig. 6.17 Joint torques, output-feedback $\mathcal{H}_\infty$ Markovian control

Table 6.6 Performance indexes—AAA–PAA–PAP fault sequence

Controller	$\mathcal{L}_2[x]$	$E[\tau]$ (N m s)
$\mathcal{H}_2$ Markovian	0.2175	0.1230
$\mathcal{H}_\infty$ Markovian	0.1936	0.1363
Mixed $\mathcal{H}_2/\mathcal{H}_\infty$ Markovian	0.1971	0.1390
Output-feedback $\mathcal{H}_\infty$ Markovian	0.2382	0.1377

The controller is computed considering $\alpha = 10$ and $\beta = 10$ for all configurations (Eq. 6.4). The best value of γ is 10. The desired positions are $q(T) = [20 - 20\ 15]^\circ$. The experimental results, including joint positions, Markovian states and joint torques are shown in Figs. 6.16 and 6.17.

Analogously to the AAA–APA fault sequence, we performed $N = 10$ experiments for each controller. The values of $\mathcal{L}_2[x]$ and $E[\tau]$ for the AAA–PAA–PAP fault sequence are shown in Table 6.6.

Appendix

State space matrices A and B and Markovian control gains K for fault sequence AAA–APA (Tables 6.7, 6.8, 6.9, 6.10, 6.11 and 6.12).

Table 6.7 Matrices A, AAA–APA fault sequence

Point	A_{AAA}						A_{APAu}						A_{APAl}					
1	1.0	−0.014	0.0049	0.057	−0.012	0.0021	1.0	0.0	0.0	0.0	0.0	0.0	1.0	0.0	−0.0056	0.052	0.0	−0.00017
	−0.028	1.0	−0.016	−0.015	0.080	−0.0068	0.0	1.1	−0.18	0.0	0.091	−0.040	0.0	1.0	0.0	0.0	0.0	0.0
	0.020	−0.032	1.0	0.010	−0.026	0.061	0.0	−0.18	1.3	0.0	−0.065	0.12	−0.018	0.0	1.1	−0.0043	0.0	0.053
	0.50	−0.56	0.20	1.3	−0.46	0.084	0.0	0.0	0.0	1.0	0.0	0.0	0.34	0.0	−0.22	1.1	0.0	−0.0070
	−1.1	1.4	−0.64	−0.59	2.2	−0.27	0.0	5.6	−7.2	0.0	2.6	−1.6	0.0	0.0	0.0	0.0	1.0	0.0
	0.79	−1.3	1.0	0.42	−1.1	1.4	0.0	−7.2	13.0	0.0	−2.6	3.9	−0.73	0.0	2.2	−0.17	0.0	1.1
2	1.0	−0.014	0.0049	0.057	−0.012	0.0021	1.0	0.0	0.0	0.0	0.0	0.0	1.0	0.0	−0.0056	0.052	0.0	−0.00017
	−0.028	1.0	−0.016	−0.015	0.080	−0.0068	0.0	1.1	−0.18	0.0	0.091	−0.040	0.0	1.0	0.0	0.0	0.0	0.0
	0.020	−0.032	1.0	0.010	−0.026	0.061	0.0	−0.18	1.3	0.0	−0.065	0.12	−0.018	0.0	1.1	−0.0043	0.0	0.053
	0.50	−0.56	0.20	1.3	−0.46	0.084	0.0	0.0	0.0	1.0	0.0	0.0	0.34	0.0	−0.22	1.1	0.0	−0.0070
	−1.1	1.4	−0.64	−0.59	2.2	−0.27	0.0	5.6	−7.2	0.0	2.6	−1.6	0.0	0.0	0.0	0.0	1.0	0.0
	0.79	−1.3	1.0	0.42	−1.1	1.4	0.0	−7.2	13.0	0.0	−2.6	3.9	−0.73	0.0	2.2	−0.17	0.0	1.1
3	1.0	−0.012	0.0043	0.056	−0.010	0.0018	1.0	0.0	0.0	0.0	0.0	0.0	1.0	0.0	−0.0055	0.052	0.0	−0.00017
	−0.024	1.0	−0.014	−0.013	0.076	−0.0061	0.0	1.1	−0.16	0.0	0.086	−0.036	0.0	1.0	0.0	0.0	0.0	0.0
	0.017	−0.029	1.0	0.0090	−0.024	0.060	0.0	−0.16	1.3	0.0	−0.059	0.12	−0.018	0.0	1.1	−0.0043	0.0	0.053
	0.44	−0.49	0.17	1.2	−0.40	0.073	0.0	0.0	0.0	1.0	0.0	0.0	0.34	0.0	−0.22	1.1	0.0	−0.0068
	−0.97	1.3	−0.58	−0.51	2.0	−0.24	0.0	4.8	−6.4	0.0	2.4	−1.4	0.0	0.0	0.0	0.0	1.0	0.0
	0.69	−1.2	0.96	0.36	−0.95	1.4	0.0	−6.3	12.0	0.0	−2.4	3.7	−0.72	0.0	2.2	−0.17	0.0	1.1
4	1.0	−0.012	0.0043	0.056	−0.010	0.0018	1.0	0.0	0.0	0.0	0.0	0.0	1.0	0.0	−0.0055	0.052	0.0	−0.00017
	−0.024	1.0	−0.014	−0.013	0.076	−0.0061	0.0	1.1	−0.16	0.0	0.086	−0.036	0.0	1.0	0.0	0.0	0.0	0.0
	0.017	−0.029	1.0	0.0090	−0.024	0.060	0.0	−0.16	1.3	0.0	−0.059	0.12	−0.018	0.0	1.1	−0.0043	0.0	0.053
	0.44	−0.49	0.17	1.2	−0.40	0.073	0.0	0.0	0.0	1.0	0.0	0.0	0.34	0.0	−0.22	1.1	0.0	−0.0068
	−0.97	1.3	−0.58	−0.51	2.0	−0.24	0.0	4.8	−6.4	0.0	2.4	−1.4	0.0	0.0	0.0	0.0	1.0	0.0
	0.69	−1.2	0.96	0.36	−0.95	1.4	0.0	−6.3	12.0	0.0	−2.4	3.7	−0.72	0.0	2.2	−0.17	0.0	1.1

Table 6.8 Matrices A, AAA–APA fault sequence

Point	A_{AAA}	$A_{\mathrm{APA}u}$	A_{APAl}
5	$\begin{bmatrix} 1.0 & -0.014 & 0.0050 & 0.057 & -0.012 & 0.0021 \\ -0.028 & 1.0 & -0.016 & -0.015 & 0.080 & -0.0067 \\ 0.020 & -0.032 & 1.0 & 0.010 & -0.026 & 0.060 \\ 0.50 & -0.56 & 0.20 & 1.3 & -0.47 & 0.085 \\ -1.1 & 1.4 & -0.63 & -0.59 & 2.2 & -0.27 \\ 0.80 & -1.3 & 0.98 & 0.42 & -1.0 & 1.4 \end{bmatrix}$	$\begin{bmatrix} 1.0 & 0.0 & 0.0 & 0.0 & 0.0 & 0.0 \\ 0.0 & 1.1 & -0.17 & 0.0 & 0.090 & -0.039 \\ 0.0 & -0.17 & 1.3 & 0.0 & -0.062 & 0.12 \\ 0.0 & 0.0 & 0.0 & 1.0 & 0.0 & 0.0 \\ 0.0 & 5.5 & -6.9 & 0.0 & 2.6 & -1.5 \\ 0.0 & -7.0 & 13.0 & 0.0 & -2.5 & 3.8 \end{bmatrix}$	$\begin{bmatrix} 1.0 & 0.0 & -0.0053 & 0.052 & 0.0 & -0.00016 \\ 0.0 & 1.0 & 0.0 & 0.0 & 0.0 & 0.0 \\ -0.017 & 0.0 & 1.1 & -0.0041 & 0.0 & 0.053 \\ 0.34 & 0.0 & -0.21 & 1.1 & 0.0 & -0.0065 \\ 0.0 & 0.0 & 0.0 & 0.0 & 1.0 & 0.0 \\ -0.69 & 0.0 & 2.2 & -0.16 & 0.0 & 1.1 \end{bmatrix}$
6	$\begin{bmatrix} 1.0 & -0.014 & 0.0050 & 0.057 & -0.012 & 0.0021 \\ -0.028 & 1.0 & -0.016 & -0.015 & 0.080 & -0.0067 \\ 0.020 & -0.032 & 1.0 & 0.010 & -0.026 & 0.060 \\ 0.50 & -0.56 & 0.20 & 1.3 & -0.47 & 0.085 \\ -1.1 & 1.4 & -0.63 & -0.59 & 2.2 & -0.27 \\ 0.80 & -1.3 & 0.98 & 0.42 & -1.0 & 1.4 \end{bmatrix}$	$\begin{bmatrix} 1.0 & 0.0 & 0.0 & 0.0 & 0.0 & 0.0 \\ 0.0 & 1.1 & -0.17 & 0.0 & 0.090 & -0.039 \\ 0.0 & -0.17 & 1.3 & 0.0 & -0.062 & 0.12 \\ 0.0 & 0.0 & 0.0 & 1.0 & 0.0 & 0.0 \\ 0.0 & 5.5 & -6.9 & 0.0 & 2.6 & -1.5 \\ 0.0 & -7.0 & 13.0 & 0.0 & -2.5 & 3.8 \end{bmatrix}$	$\begin{bmatrix} 1.0 & 0.0 & -0.0053 & 0.052 & 0.0 & -0.00016 \\ 0.0 & 1.0 & 0.0 & 0.0 & 0.0 & 0.0 \\ -0.017 & 0.0 & 1.1 & -0.0041 & 0.0 & 0.053 \\ 0.34 & 0.0 & -0.21 & 1.1 & 0.0 & -0.0065 \\ 0.0 & 0.0 & 0.0 & 0.0 & 1.0 & 0.0 \\ -0.69 & 0.0 & 2.2 & -0.16 & 0.0 & 1.1 \end{bmatrix}$
7	$\begin{bmatrix} 1.0 & -0.012 & 0.0045 & 0.056 & -0.010 & 0.0019 \\ -0.025 & 1.0 & -0.015 & -0.013 & 0.076 & -0.0063 \\ 0.018 & -0.029 & 1.0 & 0.0095 & -0.024 & 0.060 \\ 0.45 & -0.49 & 0.18 & 1.2 & -0.41 & 0.077 \\ -0.99 & 1.3 & -0.59 & -0.52 & 2.1 & -0.25 \\ 0.72 & -1.2 & 0.96 & 0.38 & -0.97 & 1.4 \end{bmatrix}$	$\begin{bmatrix} 1.0 & 0.0 & 0.0 & 0.0 & 0.0 & 0.0 \\ 0.0 & 1.1 & -0.16 & 0.0 & 0.087 & -0.036 \\ 0.0 & -0.16 & 1.3 & 0.0 & -0.060 & 0.12 \\ 0.0 & 0.0 & 0.0 & 1.0 & 0.0 & 0.0 \\ 0.0 & 5.0 & -6.5 & 0.0 & 2.5 & -1.5 \\ 0.0 & -6.5 & 12.0 & 0.0 & -2.4 & 3.7 \end{bmatrix}$	$\begin{bmatrix} 1.0 & 0.0 & -0.0051 & 0.052 & 0.0 & -0.00015 \\ 0.0 & 1.0 & 0.0 & 0.0 & 0.0 & 0.0 \\ -0.017 & 0.0 & 1.1 & -0.0039 & 0.0 & 0.053 \\ 0.34 & 0.0 & -0.20 & 1.1 & 0.0 & -0.0061 \\ 0.0 & 0.0 & 0.0 & 0.0 & 1.0 & 0.0 \\ -0.67 & 0.0 & 2.1 & -0.16 & 0.0 & 1.1 \end{bmatrix}$
8	$\begin{bmatrix} 1.0 & -0.012 & 0.0045 & 0.056 & -0.010 & 0.0019 \\ -0.025 & 1.0 & -0.015 & -0.013 & 0.076 & -0.0063 \\ 0.018 & -0.029 & 1.0 & 0.0095 & -0.024 & 0.060 \\ 0.45 & -0.49 & 0.18 & 1.2 & -0.41 & 0.077 \\ -0.99 & 1.3 & -0.59 & -0.52 & 2.1 & -0.25 \\ 0.72 & -1.2 & 0.96 & 0.38 & -0.97 & 1.4 \end{bmatrix}$	$\begin{bmatrix} 1.0 & 0.0 & 0.0 & 0.0 & 0.0 & 0.0 \\ 0.0 & 1.1 & -0.16 & 0.0 & 0.087 & -0.036 \\ 0.0 & -0.16 & 1.3 & 0.0 & -0.060 & 0.12 \\ 0.0 & 0.0 & 0.0 & 1.0 & 0.0 & 0.0 \\ 0.0 & 5.0 & -6.5 & 0.0 & 2.5 & -1.5 \\ 0.0 & -6.5 & 12.0 & 0.0 & -2.4 & 3.7 \end{bmatrix}$	$\begin{bmatrix} 1.0 & 0.0 & -0.0051 & 0.052 & 0.0 & -0.00015 \\ 0.0 & 1.0 & 0.0 & 0.0 & 0.0 & 0.0 \\ -0.017 & 0.0 & 1.1 & -0.0039 & 0.0 & 0.053 \\ 0.34 & 0.0 & -0.20 & 1.1 & 0.0 & -0.0061 \\ 0.0 & 0.0 & 0.0 & 0.0 & 1.0 & 0.0 \\ -0.67 & 0.0 & 2.1 & -0.16 & 0.0 & 1.1 \end{bmatrix}$

Table 6.9 Matrices B, AAA–APA fault sequence

Point	B_{AAA}	B_{APAu}	B_{APAl}	Point	B_{AAA}	B_{APAu}	B_{APAl}
1	$\begin{bmatrix} 1.3 & -2.8 & 2.0 \\ -2.8 & 7.2 & -6.4 \\ 2.0 & -6.4 & 10.0 \\ 50.0 & -110.0 & 79.0 \\ -110.0 & 290.0 & -260.0 \\ 79.0 & -260.0 & 400.0 \end{bmatrix}$	$\begin{bmatrix} 0.0 & 0.0 & 0.0 \\ -3.5 & 0.0 & -11.0 \\ 4.0 & 0.0 & 21.0 \\ 0.0 & 0.0 & 0.0 \\ -140.0 & 0.0 & -450.0 \\ 160.0 & 0.0 & 830.0 \end{bmatrix}$	$\begin{bmatrix} 0.16 & 0.0 & -0.49 \\ 0.0 & 0.0 & 0.0 \\ -0.50 & 0.0 & 4.2 \\ 6.5 & 0.0 & -20.0 \\ 0.0 & 0.0 & 0.0 \\ -20.0 & 0.0 & 170.0 \end{bmatrix}$	5	$\begin{bmatrix} 1.3 & -2.8 & 2.0 \\ -2.8 & 7.2 & -6.3 \\ 2.0 & -6.3 & 9.8 \\ 50.0 & -110.0 & 80.0 \\ -110.0 & 290.0 & -250.0 \\ 80.0 & -250.0 & 390.0 \end{bmatrix}$	$\begin{bmatrix} 0.0 & 0.0 & 0.0 \\ -3.5 & 0.0 & -11.0 \\ 4.0 & 0.0 & 20.0 \\ 0.0 & 0.0 & 0.0 \\ -140.0 & 0.0 & -440.0 \\ 160.0 & 0.0 & 790.0 \end{bmatrix}$	$\begin{bmatrix} 0.16 & 0.0 & -0.47 \\ 0.0 & 0.0 & 0.0 \\ -0.48 & 0.0 & 4.1 \\ 6.3 & 0.0 & -19.0 \\ 0.0 & 0.0 & 0.0 \\ -19.0 & 0.0 & 160.0 \end{bmatrix}$
2	$\begin{bmatrix} 1.3 & -2.8 & 2.0 \\ -2.8 & 7.2 & -6.4 \\ 2.0 & -6.4 & 10.0 \\ 50.0 & -110.0 & 79.0 \\ -110.0 & 290.0 & -260.0 \\ 79.0 & -260.0 & 400.0 \end{bmatrix}$	$\begin{bmatrix} 0.0 & 0.0 & 0.0 \\ -3.5 & 0.0 & -11.0 \\ 4.0 & 0.0 & 21.0 \\ 0.0 & 0.0 & 0.0 \\ -140.0 & 0.0 & -450.0 \\ 160.0 & 0.0 & 830.0 \end{bmatrix}$	$\begin{bmatrix} 0.16 & 0.0 & -0.49 \\ 0.0 & 0.0 & 0.0 \\ -0.50 & 0.0 & 4.2 \\ 6.5 & 0.0 & -20.0 \\ 0.0 & 0.0 & 0.0 \\ -20.0 & 0.0 & 170.0 \end{bmatrix}$	6	$\begin{bmatrix} 1.3 & -2.8 & 2.0 \\ -2.8 & 7.2 & -6.3 \\ 2.0 & -6.3 & 9.8 \\ 50.0 & -110.0 & 80.0 \\ -110.0 & 290.0 & -250.0 \\ 80.0 & -250.0 & 390.0 \end{bmatrix}$	$\begin{bmatrix} 0.0 & 0.0 & 0.0 \\ -3.5 & 0.0 & -11.0 \\ 4.0 & 0.0 & 20.0 \\ 0.0 & 0.0 & 0.0 \\ -140.0 & 0.0 & -440.0 \\ 160.0 & 0.0 & 790.0 \end{bmatrix}$	$\begin{bmatrix} 0.16 & 0.0 & -0.47 \\ 0.0 & 0.0 & 0.0 \\ -0.48 & 0.0 & 4.1 \\ 6.3 & 0.0 & -19.0 \\ 0.0 & 0.0 & 0.0 \\ -19.0 & 0.0 & 160.0 \end{bmatrix}$
3	$\begin{bmatrix} 1.1 & -2.4 & 1.7 \\ -2.4 & 6.3 & -5.8 \\ 1.7 & -5.8 & 9.6 \\ 44.0 & -97.0 & 69.0 \\ -97.0 & 250.0 & -230.0 \\ 69.0 & -230.0 & 380.0 \end{bmatrix}$	$\begin{bmatrix} 0.0 & 0.0 & 0.0 \\ -3.0 & 0.0 & -10.0 \\ 3.5 & 0.0 & 19.0 \\ 0.0 & 0.0 & 0.0 \\ -120.0 & 0.0 & -400.0 \\ 140.0 & 0.0 & 770.0 \end{bmatrix}$	$\begin{bmatrix} 0.16 & 0.0 & -0.49 \\ 0.0 & 0.0 & 0.0 \\ -0.49 & 0.0 & 4.2 \\ 6.5 & 0.0 & -19.0 \\ 0.0 & 0.0 & 0.0 \\ -20.0 & 0.0 & 170.0 \end{bmatrix}$	7	$\begin{bmatrix} 1.1 & -2.5 & 1.8 \\ -2.5 & 6.4 & -5.9 \\ 1.8 & -5.9 & 9.6 \\ 45.0 & -99.0 & 72.0 \\ -99.0 & 260.0 & -240.0 \\ 72.0 & -240.0 & 380.0 \end{bmatrix}$	$\begin{bmatrix} 0.0 & 0.0 & 0.0 \\ -3.1 & 0.0 & -10.0 \\ 3.6 & 0.0 & 19.0 \\ 0.0 & 0.0 & 0.0 \\ -120.0 & 0.0 & -410.0 \\ 150.0 & 0.0 & 770.0 \end{bmatrix}$	$\begin{bmatrix} 0.16 & 0.0 & -0.46 \\ 0.0 & 0.0 & 0.0 \\ -0.46 & 0.0 & 4.0 \\ 6.3 & 0.0 & -18.0 \\ 0.0 & 0.0 & 0.0 \\ -18.0 & 0.0 & 160.0 \end{bmatrix}$
4	$\begin{bmatrix} 1.1 & -2.4 & 1.7 \\ -2.4 & 6.3 & -5.8 \\ 1.7 & -5.8 & 9.6 \\ 44.0 & -97.0 & 69.0 \\ -97.0 & 250.0 & -230.0 \\ 69.0 & -230.0 & 380.0 \end{bmatrix}$	$\begin{bmatrix} 0.0 & 0.0 & 0.0 \\ -3.0 & 0.0 & -10.0 \\ 3.5 & 0.0 & 19.0 \\ 0.0 & 0.0 & 0.0 \\ -120.0 & 0.0 & -400.0 \\ 140.0 & 0.0 & 770.0 \end{bmatrix}$	$\begin{bmatrix} 0.16 & 0.0 & -0.49 \\ 0.0 & 0.0 & 0.0 \\ -0.49 & 0.0 & 4.2 \\ 6.5 & 0.0 & -19.0 \\ 0.0 & 0.0 & 0.0 \\ -20.0 & 0.0 & 170.0 \end{bmatrix}$	8	$\begin{bmatrix} 1.1 & -2.5 & 1.8 \\ -2.5 & 6.4 & -5.9 \\ 1.8 & -5.9 & 9.6 \\ 45.0 & -99.0 & 72.0 \\ -99.0 & 260.0 & -240.0 \\ 72.0 & -240.0 & 380.0 \end{bmatrix}$	$\begin{bmatrix} 0.0 & 0.0 & 0.0 \\ -3.1 & 0.0 & -10.0 \\ 3.6 & 0.0 & 19.0 \\ 0.0 & 0.0 & 0.0 \\ -120.0 & 0.0 & -410.0 \\ 150.0 & 0.0 & 770.0 \end{bmatrix}$	$\begin{bmatrix} 0.16 & 0.0 & -0.46 \\ 0.0 & 0.0 & 0.0 \\ -0.46 & 0.0 & 4.0 \\ 6.3 & 0.0 & -18.0 \\ 0.0 & 0.0 & 0.0 \\ -18.0 & 0.0 & 160.0 \end{bmatrix}$

Table 6.10 H_2 controller gains, AAA–APA fault sequence

Point	K_{AAA}	K_{APAu}	K_{APAl}
1	$\begin{bmatrix} -0.24 & -0.12 & -0.026 & -0.24 & -0.11 & -0.027 \\ -0.11 & -0.072 & -0.018 & -0.11 & -0.070 & -0.018 \\ -0.027 & -0.019 & -0.011 & -0.027 & -0.018 & -0.010 \end{bmatrix}$	$\begin{bmatrix} 0.0 & 0.052 & 0.012 & 0.0 & 0.024 & 0.011 \\ 0.0 & 0.0 & 0.0 & 0.0 & 0.0 & 0.0 \\ -0.0 & -0.0015 & -0.019 & -0.0 & -0.0016 & -0.0069 \end{bmatrix}$	$\begin{bmatrix} -0.29 & -0.0 & -0.036 & -0.25 & -0.0 & -0.029 \\ 0.0 & 0.0 & 0.0 & 0.0 & 0.0 & 0.0 \\ -0.030 & -0.0 & -0.023 & -0.029 & -0.0 & -0.010 \end{bmatrix}$
2	$\begin{bmatrix} -0.24 & -0.12 & -0.026 & -0.24 & -0.11 & -0.027 \\ -0.11 & -0.072 & -0.018 & -0.11 & -0.070 & -0.018 \\ -0.027 & -0.019 & -0.011 & -0.027 & -0.018 & -0.010 \end{bmatrix}$	$\begin{bmatrix} 0.0 & 0.052 & 0.012 & 0.0 & 0.024 & 0.011 \\ 0.0 & 0.0 & 0.0 & 0.0 & 0.0 & 0.0 \\ -0.05 & -0.0015 & -0.019 & -0.0 & -0.0016 & -0.0069 \end{bmatrix}$	$\begin{bmatrix} -0.29 & -0.0 & -0.036 & -0.25 & -0.0 & -0.029 \\ 0.0 & 0.0 & 0.0 & 0.0 & 0.0 & 0.0 \\ -0.030 & -0.0 & -0.023 & -0.029 & -0.0 & -0.010 \end{bmatrix}$
3	$\begin{bmatrix} -0.24 & -0.11 & -0.026 & -0.24 & -0.11 & -0.027 \\ -0.11 & -0.073 & -0.018 & -0.11 & -0.070 & -0.018 \\ -0.026 & -0.019 & -0.011 & -0.027 & -0.019 & -0.010 \end{bmatrix}$	$\begin{bmatrix} 0.0 & 0.054 & 0.012 & 0.0 & 0.026 & 0.011 \\ 0.0 & 0.0 & 0.0 & 0.0 & 0.0 & 0.0 \\ -0.0 & -0.0015 & -0.019 & -0.0 & -0.0016 & -0.0069 \end{bmatrix}$	$\begin{bmatrix} -0.28 & -0.0 & -0.035 & -0.25 & -0.0 & -0.029 \\ 0.0 & 0.0 & 0.0 & 0.0 & 0.0 & 0.0 \\ -0.029 & -0.0 & -0.023 & -0.028 & -0.0 & -0.010 \end{bmatrix}$
4	$\begin{bmatrix} -0.24 & -0.11 & -0.026 & -0.240 & -0.11 & -0.027 \\ -0.11 & -0.073 & -0.018 & -0.11 & -0.070 & -0.018 \\ -0.026 & -0.019 & -0.011 & -0.027 & -0.019 & -0.010 \end{bmatrix}$	$\begin{bmatrix} 0.0 & 0.054 & 0.012 & 0.0 & 0.026 & 0.011 \\ 0.0 & 0.0 & 0.0 & 0.0 & 0.0 & 0.0 \\ -0.0 & -0.0015 & -0.019 & -0.0 & -0.0016 & -0.0069 \end{bmatrix}$	$\begin{bmatrix} -0.28 & -0.0 & -0.035 & -0.25 & -0.0 & -0.029 \\ 0.0 & 0.0 & 0.0 & 0.0 & 0.0 & 0.0 \\ -0.029 & -0.0 & -0.023 & -0.028 & -0.0 & -0.010 \end{bmatrix}$
5	$\begin{bmatrix} -0.24 & -0.12 & -0.025 & -0.24 & -0.11 & -0.027 \\ -0.11 & -0.072 & -0.017 & -0.11 & -0.069 & -0.018 \\ -0.026 & -0.019 & -0.011 & -0.027 & -0.018 & -0.010 \end{bmatrix}$	$\begin{bmatrix} 0.0 & 0.052 & 0.012 & 0.0 & 0.024 & 0.011 \\ 0.0 & 0.0 & 0.0 & 0.0 & 0.0 & 0.0 \\ -0.0 & -0.0017 & -0.019 & -0.0 & -0.0018 & -0.0070 \end{bmatrix}$	$\begin{bmatrix} -0.29 & -0.0 & -0.035 & -0.25 & -0.0 & -0.028 \\ 0.0 & 0.0 & 0.0 & 0.0 & 0.0 & 0.0 \\ -0.029 & -0.0 & -0.023 & -0.028 & -0.0 & -0.010 \end{bmatrix}$
6	$\begin{bmatrix} -0.24 & -0.12 & -0.025 & -0.24 & -0.11 & -0.027 \\ -0.11 & -0.072 & -0.017 & -0.11 & -0.069 & -0.018 \\ -0.026 & -0.019 & -0.011 & -0.027 & -0.018 & -0.010 \end{bmatrix}$	$\begin{bmatrix} 0.0 & 0.052 & 0.012 & 0.0 & 0.024 & 0.011 \\ 0.0 & 0.0 & 0.0 & 0.0 & 0.0 & 0.0 \\ -0.0 & -0.0017 & -0.019 & -0.0 & -0.0018 & -0.0070 \end{bmatrix}$	$\begin{bmatrix} -0.29 & -0.0 & -0.035 & -0.255 & -0.0 & -0.028 \\ 0.0 & 0.0 & 0.0 & 0.0 & 0.0 & 0.0 \\ -0.029 & -0.0 & -0.023 & -0.028 & -0.0 & -0.010 \end{bmatrix}$
7	$\begin{bmatrix} -0.23 & -0.11 & -0.025 & -0.23 & -0.11 & -0.026 \\ -0.11 & -0.072 & -0.017 & -0.11 & -0.069 & -0.018 \\ -0.025 & -0.019 & -0.011 & -0.026 & -0.018 & -0.010 \end{bmatrix}$	$\begin{bmatrix} 0.0 & 0.054 & 0.013 & 0.0 & 0.026 & 0.012 \\ 0.0 & 0.0 & 0.0 & 0.0 & 0.0 & 0.0 \\ -0.0 & -0.0018 & -0.019 & -0.0 & -0.0019 & -0.0071 \end{bmatrix}$	$\begin{bmatrix} -0.28 & -0.0 & -0.034 & -0.25 & -0.0 & -0.028 \\ 0.0 & 0.0 & 0.0 & 0.0 & 0.0 & 0.0 \\ -0.028 & -0.0 & -0.023 & -0.027 & -0.0 & -0.010 \end{bmatrix}$
8	$\begin{bmatrix} -0.23 & -0.11 & -0.025 & -0.23 & -0.11 & -0.026 \\ -0.11 & -0.072 & -0.017 & -0.11 & -0.069 & -0.018 \\ -0.025 & -0.019 & -0.011 & -0.026 & -0.018 & -0.010 \end{bmatrix}$	$\begin{bmatrix} 0.0 & 0.054 & 0.013 & 0.0 & 0.026 & 0.012 \\ 0.0 & 0.0 & 0.0 & 0.0 & 0.0 & 0.0 \\ -0.0 & -0.0018 & -0.019 & -0.0 & -0.0019 & -0.0071 \end{bmatrix}$	$\begin{bmatrix} -0.28 & -0.0 & -0.034 & -0.25 & -0.0 & -0.028 \\ 0.0 & 0.0 & 0.0 & 0.0 & 0.0 & 0.0 \\ -0.028 & -0.0 & -0.023 & -0.027 & -0.0 & -0.010 \end{bmatrix}$

Table 6.11 H_∞ controller gains, AAA–APA fault sequence ($\times 10^2$)

Point	K_{AAA}	K_{APAu}	K_{APAl}
1	$\begin{bmatrix} -5.88 & -1.91 & 0.204 & -4.17 & -1.65 & -0.319 \\ -0.727 & -2.79 & -1.46 & -1.34 & -1.53 & -0.443 \\ 0.200 & -0.390 & -2.69 & -0.191 & -0.287 & -0.516 \end{bmatrix}$	$\begin{bmatrix} 0.210 & 7.27 & 2.55 & 0.096 & 1.45 & 0.594 \\ 0.0 & 0.0 & 0.0 & 0.0 & 0.0 & 0.0 \\ -0.089 & -0.137 & -2.74 & -0.041 & 0.081 & -0.568 \end{bmatrix}$	$\begin{bmatrix} -0.141 & -0.016 & -1.68 & -7.11 & -0.00 & -0.785 \\ 0.0 & 0.0 & 0.0 & 0.0 & 0.0 & 0.0 \\ -0.387 & -0.097 & -4.41 & -0.585 & -0.0141 & -0.618 \end{bmatrix}$
2	$\begin{bmatrix} -5.87 & -1.90 & 0.206 & -4.15 & -1.64 & -0.318 \\ -0.730 & -2.79 & -1.46 & -1.34 & -1.53 & -0.443 \\ 0.197 & -0.387 & -2.69 & -0.192 & -0.287 & -0.516 \end{bmatrix}$	$\begin{bmatrix} 0.211 & 7.27 & 2.56 & 0.0966 & 1.46 & 0.597 \\ 0.0 & 0.0 & 0.0 & 0.0 & 0.0 & 0.0 \\ -0.089 & -0.136 & -2.74 & -0.0411 & 0.081 & -0.568 \end{bmatrix}$	$\begin{bmatrix} -0.141 & -0.016 & -1.68 & -7.10 & -0.0 & -0.783 \\ 0.0 & 0.0 & 0.0 & 0.0 & 0.0 & 0.0 \\ -0.383 & -0.097 & -4.41 & -0.583 & -0.014 & -0.618 \end{bmatrix}$
3	$\begin{bmatrix} -5.93 & -1.86 & 0.194 & -4.17 & -1.64 & -0.318 \\ -0.634 & -2.90 & -1.44 & -1.31 & -1.55 & -0.446 \\ 0.185 & -0.380 & -2.70 & -0.193 & -0.289 & -0.516 \end{bmatrix}$	$\begin{bmatrix} 0.207 & 7.38 & 2.52 & 0.0945 & 1.51 & 0.599 \\ 0.0 & 0.0 & 0.0 & 0.0 & 0.0 & 0.0 \\ -0.089 & -0.058 & -2.75 & -0.041 & 0.094 & -0.568 \end{bmatrix}$	$\begin{bmatrix} -0.142 & -0.016 & -1.68 & -7.10 & -0.0 & -0.782 \\ 0.0 & 0.0 & 0.0 & 0.0 & 0.0 & 0.0 \\ -0.389 & -0.103 & -4.42 & -0.584 & -0.0155 & -0.620 \end{bmatrix}$
4	$\begin{bmatrix} -5.93 & -1.86 & 0.188 & -4.17 & -1.63 & -0.317 \\ -0.624 & -2.90 & -1.44 & -1.30 & -1.55 & -0.445 \\ 0.201 & -0.373 & -2.70 & -0.181 & -0.283 & -0.515 \end{bmatrix}$	$\begin{bmatrix} 0.207 & 7.39 & 2.52 & 0.094 & 1.51 & 0.600 \\ 0.0 & 0.0 & 0.0 & 0.0 & 0.0 & 0.0 \\ -0.089 & -0.060 & -2.75 & -0.0412 & 0.093 & -0.568 \end{bmatrix}$	$\begin{bmatrix} -0.142 & -0.017 & -1.69 & -7.09 & -0.0 & -0.781 \\ 0.0 & 0.0 & 0.0 & 0.0 & 0.0 & 0.0 \\ -0.350 & -0.103 & -4.42 & -0.563 & -0.015 & -0.618 \end{bmatrix}$
5	$\begin{bmatrix} -5.87 & -1.90 & 0.227 & -4.15 & -1.64 & -0.308 \\ -0.743 & -2.80 & -1.47 & -1.34 & -1.53 & -0.443 \\ 0.185 & -0.393 & -2.72 & -0.195 & -0.290 & -0.522 \end{bmatrix}$	$\begin{bmatrix} 0.209 & 7.25 & 2.58 & 0.0952 & 1.46 & 0.603 \\ 0.0 & 0.0 & 0.0 & 0.0 & 0.0 & 0.0 \\ -0.090 & -0.123 & -2.78 & -0.0416 & 0.079 & -0.576 \end{bmatrix}$	$\begin{bmatrix} -0.141 & -0.015 & -1.64 & -7.09 & -0.0 & -0.767 \\ 0.0 & 0.0 & 0.0 & 0.0 & 0.0 & 0.0 \\ -0.436 & -0.099 & -4.44 & -0.601 & -0.0146 & -0.626 \end{bmatrix}$
6	$\begin{bmatrix} -5.87 & -1.90 & 0.220 & -4.14 & -1.63 & -0.307 \\ -0.731 & -2.79 & -1.47 & -1.33 & -1.52 & -0.442 \\ 0.206 & -0.384 & -2.72 & -0.179 & -0.283 & -0.521 \end{bmatrix}$	$\begin{bmatrix} 0.209 & 7.26 & 2.57 & 0.0953 & 1.46 & 0.605 \\ 0.0 & 0.0 & 0.0 & 0.0 & 0.0 & 0.0 \\ -0.090 & -0.126 & -2.78 & -0.041 & 0.078 & -0.576 \end{bmatrix}$	$\begin{bmatrix} -0.141 & -0.016 & -1.65 & -7.08 & -0.0 & -0.766 \\ 0.0 & 0.0 & 0.0 & 0.0 & 0.0 & 0.0 \\ -0.381 & -0.099 & -4.44 & -0.573 & -0.0145 & -0.623 \end{bmatrix}$
7	$\begin{bmatrix} -5.94 & -1.86 & 0.229 & -4.16 & -1.63 & -0.301 \\ -0.64 & -2.91 & -1.47 & -1.30 & -1.55 & -0.449 \\ 0.193 & -0.394 & -2.71 & -0.183 & -0.290 & -0.521 \end{bmatrix}$	$\begin{bmatrix} 0.205 & 7.35 & 2.57 & 0.093 & 1.52 & 0.616 \\ 0.0 & 0.0 & 0.0 & 0.0 & 0.0 & 0.0 \\ -0.0881 & -0.073 & -2.77 & -0.04 & 0.085 & -0.574 \end{bmatrix}$	$\begin{bmatrix} -0.142 & -0.015 & -1.63 & -7.08 & -0 & -0.759 \\ 0.0 & 0.0 & 0.0 & 0.0 & 0.0 & 0.0 \\ -0.424 & -0.108 & -4.46 & -0.589 & -0.016 & -0.629 \end{bmatrix}$
8	$\begin{bmatrix} -5.93 & -1.85 & 0.229 & -4.15 & -1.62 & -0.300 \\ -0.642 & -2.90 & -1.47 & -1.30 & -1.55 & -0.449 \\ 0.195 & -0.389 & -2.71 & -0.179 & -0.289 & -0.521 \end{bmatrix}$	$\begin{bmatrix} 0.205 & 7.36 & 2.58 & 0.093 & 1.52 & 0.620 \\ 0.0 & 0.0 & 0.0 & 0.0 & 0.0 & 0.0 \\ -0.088 & -0.073 & -2.77 & -0.04 & 0.084 & -0.575 \end{bmatrix}$	$\begin{bmatrix} -0.142 & -0.0157 & -1.63 & -7.07 & -0.0 & -0.757 \\ 0.0 & 0.0 & 0.0 & 0.0 & 0.0 & 0.0 \\ -0.405 & -0.107 & -4.46 & -0.578 & -0.016 & -0.628 \end{bmatrix}$

Table 6.12 H_2/H_∞ controller gains, AAA–APA fault sequence

Point	K_{AAA}	$K_{\mathrm{APA}u}$	K_{APAl}
1	$\begin{bmatrix} -1.23 & -0.33 & 0.085 & -0.87 & -0.31 & -0.058 \\ 0.061 & -0.75 & -0.32 & -0.18 & -0.33 & -0.094 \\ -0.044 & 0.066 & -0.79 & -0.053 & -0.042 & -0.12 \end{bmatrix}$	$\begin{bmatrix} 0.11 & 1.53 & 0.48 & 0.058 & 0.31 & 0.11 \\ 0.0 & 0.0 & 0.0 & 0.0 & 0.0 & 0.0 \\ -0.12 & 0.23 & -0.87 & -0.065 & 0.059 & -0.15 \end{bmatrix}$	$\begin{bmatrix} -2.69 & -0.012 & -0.32 & -1.42 & -0.0024 & -0.15 \\ 0.0 & 0.0 & 0.0 & 0.0 & 0.0 & 0.0 \\ 0.056 & -0.057 & -1.10 & -0.042 & -0.010 & -0.14 \end{bmatrix}$
2	$\begin{bmatrix} -1.22 & -0.33 & 0.086 & -0.87 & -0.31 & -0.057 \\ 0.059 & -0.75 & -0.32 & -0.18 & -0.33 & -0.095 \\ -0.044 & 0.067 & -0.78 & -0.052 & -0.042 & -0.11 \end{bmatrix}$	$\begin{bmatrix} 0.11 & 1.53 & 0.48 & 0.058 & 0.31 & 0.12 \\ 0.0 & 0.0 & 0.0 & 0.0 & 0.0 & 0.0 \\ -0.12 & 0.23 & -0.87 & -0.065 & 0.060 & -0.15 \end{bmatrix}$	$\begin{bmatrix} -2.69 & -0.012 & -0.32 & -1.42 & -0.0024 & -0.15 \\ 0.0 & 0.0 & 0.0 & 0.0 & 0.0 & 0.0 \\ 0.057 & -0.057 & -1.10 & -0.041 & -0.010 & -0.14 \end{bmatrix}$
3	$\begin{bmatrix} -1.24 & -0.33 & 0.080 & -0.88 & -0.31 & -0.058 \\ 0.075 & -0.77 & -0.31 & -0.17 & -0.33 & -0.094 \\ -0.044 & 0.065 & -0.79 & -0.053 & -0.043 & -0.12 \end{bmatrix}$	$\begin{bmatrix} 0.11 & 1.52 & 0.46 & 0.055 & 0.32 & 0.11 \\ 0.0 & 0.0 & 0.0 & 0.0 & 0.0 & 0.0 \\ -0.12 & 0.23 & -0.88 & -0.065 & 0.06 & -0.151 \end{bmatrix}$	$\begin{bmatrix} -2.71 & -0.012 & -0.32 & -1.42 & -0.0024 & -0.15 \\ 0.0 & 0.0 & 0.0 & 0.0 & 0.0 & 0.0 \\ 0.054 & -0.059 & -1.10 & -0.042 & -0.011 & -0.14 \end{bmatrix}$
4	$\begin{bmatrix} -1.22 & -0.33 & 0.090 & -0.87 & -0.31 & -0.055 \\ 0.055 & -0.76 & -0.32 & -0.18 & -0.33 & -0.095 \\ -0.056 & 0.062 & -0.79 & -0.060 & -0.045 & -0.12 \end{bmatrix}$	$\begin{bmatrix} 0.11 & 1.52 & 0.47 & 0.055 & 0.32 & 0.11 \\ 0.0 & 0.0 & 0.0 & 0.0 & 0.0 & 0.0 \\ -0.12 & 0.23 & -0.88 & -0.065 & 0.059 & -0.15 \end{bmatrix}$	$\begin{bmatrix} -2.71 & -0.012 & -0.32 & -1.42 & -0.0024 & -0.15 \\ 0.0 & 0.0 & 0.0 & 0.0 & 0.0 & 0.0 \\ 0.075 & -0.059 & -1.10 & -0.031 & -0.012 & -0.14 \end{bmatrix}$
5	$\begin{bmatrix} -1.22 & -0.33 & 0.090 & -0.87 & -0.31 & -0.055 \\ 0.055 & -0.76 & -0.32 & -0.18 & -0.33 & -0.095 \\ -0.056 & 0.062 & -0.79 & -0.060 & -0.045 & -0.12 \end{bmatrix}$	$\begin{bmatrix} 0.11 & 1.53 & 0.49 & 0.057 & 0.31 & 0.12 \\ 0.0 & 0.0 & 0.0 & 0.0 & 0.0 & 0.0 \\ -0.12 & 0.24 & -0.88 & -0.064 & 0.060 & -0.15 \end{bmatrix}$	$\begin{bmatrix} -2.69 & -0.012 & -0.31 & -1.42 & -0.0023 & -0.14 \\ 0.0 & 0.0 & 0.0 & 0.0 & 0.0 & 0.0 \\ 0.029 & -0.058 & -1.11 & -0.055 & -0.011 & -0.14 \end{bmatrix}$
6	$\begin{bmatrix} -1.22 & -0.33 & 0.089 & -0.87 & -0.31 & -0.055 \\ 0.057 & -0.75 & -0.32 & -0.18 & -0.33 & -0.095 \\ -0.042 & 0.068 & -0.79 & -0.049 & -0.041 & -0.120 \end{bmatrix}$	$\begin{bmatrix} 0.11 & 1.54 & 0.49 & 0.057 & 0.32 & 0.12 \\ 0.0 & 0.0 & 0.0 & 0.0 & 0.0 & 0.0 \\ -0.12 & 0.23 & -0.88 & -0.064 & 0.059 & -0.15 \end{bmatrix}$	$\begin{bmatrix} -2.69 & -0.012 & -0.31 & -1.41 & -0.0024 & -0.14 \\ 0.0 & 0.0 & 0.0 & 0.0 & 0.0 & 0.0 \\ 0.057 & -0.058 & -1.10 & -0.039 & -0.011 & -0.14 \end{bmatrix}$
7	$\begin{bmatrix} -1.24 & -0.33 & 0.087 & -0.88 & -0.31 & -0.055 \\ 0.072 & -0.77 & -0.31 & -0.17 & -0.33 & -0.095 \\ -0.053 & 0.061 & -0.79 & -0.056 & -0.045 & -0.12 \end{bmatrix}$	$\begin{bmatrix} 0.10 & 1.52 & 0.47 & 0.054 & 0.32 & 0.12 \\ 0.0 & 0.0 & 0.0 & 0.0 & 0.0 & 0.0 \\ -0.12 & 0.24 & -0.88 & -0.063 & 0.060 & -0.15 \end{bmatrix}$	$\begin{bmatrix} -2.71 & -0.012 & -0.31 & -1.42 & -0.0023 & -0.14 \\ 0.0 & 0.0 & 0.0 & 0.0 & 0.0 & 0.0 \\ 0.034 & -0.061 & -1.11 & -0.050 & -0.011 & -0.14 \end{bmatrix}$
8	$\begin{bmatrix} -1.24 & -0.33 & 0.088 & -0.88 & -0.31 & -0.054 \\ 0.070 & -0.77 & -0.32 & -0.17 & -0.34 & -0.096 \\ -0.048 & 0.065 & -0.79 & -0.052 & -0.043 & -0.12 \end{bmatrix}$	$\begin{bmatrix} 0.11 & 1.53 & 0.48 & 0.055 & 0.32 & 0.12 \\ 0.0 & 0.0 & 0.0 & 0.0 & 0.0 & 0.0 \\ -0.12 & 0.24 & -0.88 & -0.062 & 0.060 & -0.15 \end{bmatrix}$	$\begin{bmatrix} -2.71 & -0.012 & -0.31 & -1.42 & -0.0023 & -0.14 \\ 0.0 & 0.0 & 0.0 & 0.0 & 0.0 & 0.0 \\ 0.043 & -0.062 & -1.11 & -0.046 & -0.012 & -0.14 \end{bmatrix}$

References

1. Arai H, Tachi S (1991) Position control of a manipulator with passive joints using dynamic coupling. IEEE Trans Rob Autom 7(4):528–534
2. Bergerman M, Xu Y (1996) Optimal control sequence for underactuated manipulators. In: Proceedings of the IEEE international conference on robotics and automation, Minneapolis
3. Bergerman M, Xu Y (1996) Robust joint and cartesian control of underactuated manipulators. Trans ASME J Dyn Syst Meas Control 118(3):557–565
4. Costa OLV, do Val JBR (1996) Full information $\mathcal{H}_\infty$ control for discrete-time infinite Markov jump parameter systems. J Math Anal Appl 202:578–603
5. Costa OLV, Fragoso MD, Marques RP (2005) Discrete-time Markov jump linear systems. Springer-Verlag, London
6. Costa OLV, Marques RP (1998) Mixed $\mathcal{H}_2/\mathcal{H}_\infty$-control of discrete-time Markovian jump linear systems. IEEE Trans Autom Control 43(1):95–100
7. Costa OLV, Marques RP (2000) Robust $\mathcal{H}_2$-control for discrete-time Markovian jump linear systems. Int J Control 73(1):11–21
8. de Farias DP, Geromel JC, do Val JBR, Costa OLV (2000) Output feedback control of Markov jump linear systems in continuous-time. IEEE Trans Autom Control 45(5):944–949
9. Dixon WE, Walker ID, Dawson DM, Hartranft JP (2000) Fault detection for robot manipulators with parametric uncertainty: a prediction-error-based approach. IEEE Trans Rob Autom 16(6):689–699
10. Siqueira AAG, Terra MH (2004) A fault tolerant manipulator robot based on $\mathcal{H}_2$, $\mathcal{H}_\infty$, and mixed $\mathcal{H}_2/\mathcal{H}_\infty$ Markovian controls. In: Proceedings of the 2004 IEEE international conference on control applications (CCA), Taipei, Taiwan, pp 309–314
11. Siqueira AAG, Terra MH (2004b) Nonlinear and Markovian $\mathcal{H}_\infty$ controls of underactuated manipulators. IEEE Trans Control Syst Technol 12(6):811–826
12. Siqueira AAG, Terra MH (2009). A fault tolerant manipulator robot based on $\mathcal{H}_2$, $\mathcal{H}_\infty$, and mixed $\mathcal{H}_2/\mathcal{H}_\infty$ Markovian jump controls. IEEE/ASME Trans Mechatron 14(2):257–263

Part III
Cooperative Robot Manipulators

Chapter 7
Underactuated Cooperative Manipulators

7.1 Introduction

Cooperative manipulators offer significant advantages over single manipulators when operating on heavy, large, or flexible loads [1, 4, 5, 7, 9–11, 13, 15]. As in humans, robotic systems with two manipulators can execute tasks that are difficult or even impossible to be performed by a single robot (e.g., assembly of large structures) [11].

Control of cooperative manipulators is complex because of the interaction between the arms caused by the kinematic and internal force constraints. Position control must be coordinated with control of the squeeze forces in the load to avoid damaging it. Several solutions have been proposed in the literature to deal with this problem in fault-free cooperative manipulators rigidly connected to an undeformable load. They include the master/slave strategy [8], the optimal division of the load control [4, 9], the definition of new task objectives or variables [3, 6], and the hybrid control of motion and squeeze forces [2, 10, 14]. Despite presenting good performance for the fault-free case, these controllers, in general, were not designed to account for the presence of passive joints.

This chapter presents solutions to three important problems related to cooperative manipulators:

1. A stable motion control method for underactuated cooperative manipulator systems with any number of manipulators (Sects. 7.2 and 7.3). The control of several cooperative manipulators with passive joints, instead of only two as in [7], is made possible by the introduction of a new Jacobian matrix $Q(q)$.
2. An extension of the squeeze force control in [14] to cooperative manipulators with passive joints (Sect. 7.3). The proposed strategy is of a hybrid nature and treats the load's motion and the components of the squeeze forces independently.
3. A method to measure the dynamic load carrying capacity of cooperative manipulators with passive joints (Sect. 7.4).

A. A. G. Siqueira et al., *Robust Control of Robots*,
DOI: 10.1007/978-0-85729-898-0_7, © Springer-Verlag London Limited 2011

Section 7.5 presents the results of the application of these methods to the UARM.

7.2 Cooperative Manipulators

In this section, we develop the complete kinematic and dynamic model of cooperative manipulators, giving a special attention to the interaction forces between the load and the manipulators' end-effectors.

Consider a robotic system with m manipulators rigidly connected to an undeformable load (see Fig. 7.1, where the case $m = 2$ is illustrated for clarity of presentation). Let q_i be the vector of generalized coordinates of manipulator i and $x_o = [p_o^T \ \phi_o^T]^T$ be the k-dimensional vector of load position and orientation. In the three-dimensional space, $p_o = [x_o \ y_o \ z_o]^T$ is the position of the origin of the frame attached to the center of mass of the load (frame CM) with respect to an appropriately selected origin (e.g., the base of one of the manipulators), and $\phi_o = [\varphi_o \ \upsilon_o \ \psi_o]^T$ is a minimal representation of the load orientation using, say, Euler angles or RPY (roll-pitch-yaw) angles.

As it is possible to compute the positions and orientations of the load using the positions of the joints of any arm, the cooperative system presents the following kinematic constraints:

$$x_o = \phi_i(q_i), \tag{7.1}$$

for $i = 1, 2, \ldots, m$, where $\phi_i(q_i)$ is the representation of the direct kinematics of manipulator i. Differentiating Eq. 7.1 with respect to time, the following velocity constraint appears:

$$\dot{x}_o = D_i(x_o, q_i)\dot{q}_i, \tag{7.2}$$

for $i = 1, 2, \ldots, m$, where $D_i(x_o, q_i) = J_{o_i}^{-1}(x_o)J_i(q_i)$ is the Jacobian matrix relating joint velocities of manipulator i and load velocities, $J_i(q_i)$ is the Jacobian matrix of manipulator i (from joint velocity to end-effector velocity) and $J_{o_i}(x_o)$ is the Jacobian matrix that converts velocities of the load into velocities of the end-effector of arm i. Since the load position and orientation can be computed through the joint positions of the manipulators and vice versa, the above Jacobian matrices are represented as $D_i(q_i)$, $J_i(q_i)$ and $J_{o_i}(x_o)$.

The equation of motion for the ith manipulator in the cooperative system is given by:

$$M_i(q_i)\ddot{q}_i + C_i(q_i, \dot{q}_i)\dot{q}_i + g_i(q_i) = \tau_i - J_i(q_i)^T h_i, \tag{7.3}$$

where $M_i(q_i)$ is the inertia matrix, $C_i(q_i, \dot{q}_i)$ is the matrix of centrifugal and Coriolis terms, $g_i(q_i)$ is the vector of gravitational torques, τ_i is the vector of applied torques, and $h_i = [f_i^T \ \eta_i^T]^T$ is the force vector at the end-effector, with f_i

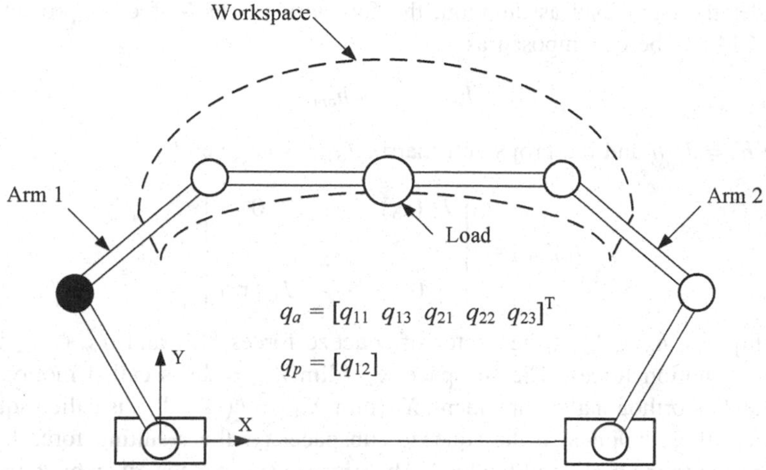

Fig. 7.1 Two 3-joint cooperative manipulators handling a common load. In this example the second joint of the manipulator on the left is passive

representing linear forces and η_i representing torques; the friction terms are not shown for clarity.

The combined dynamics of all manipulators can be written as:

$$M(q)\ddot{q} + C(q,\dot{q})\dot{q} + g(q) = \tau - J(q)^T h, \tag{7.4}$$

where $q = [q_1^T\ q_2^T \ldots q_m^T]^T$, $\tau = [\tau_1^T\ \tau_2^T \ldots \tau_m^T]^T$, $h = [h_1^T\ h_2^T \ldots h_m^T]^T$, $g = [g_1^T\ g_2^T \cdots g_m^T]^T$,

$$M(q) = \begin{bmatrix} M_1(q_1) & \ldots & 0 \\ \vdots & \ddots & \vdots \\ 0 & \ldots & M_m(q_m) \end{bmatrix},$$

$$C(q,\dot{q}) = \begin{bmatrix} C_1(q_1,\dot{q}_1) & \ldots & 0 \\ \vdots & \ddots & \vdots \\ 0 & \ldots & C_m(q_m,\dot{q}_m) \end{bmatrix},$$

and $$J(q) = \begin{bmatrix} J_1(q_1) & \cdots & 0 \\ \vdots & \ddots & \vdots \\ 0 & \cdots & J_m(q_m) \end{bmatrix}.$$

The equation of motion of the load is given by:

$$M_o\ddot{x}_o + b_o(x_o,\dot{x}_o) = J_o(x_o)^T h, \tag{7.5}$$

where M_o is the load inertia matrix, $b_o(x_o,\dot{x}_o)$ is the vector of centrifugal, Coriolis, and gravitational torques, and $J_o(x_o) = [J_{o_1}(x_o)^T, \ldots, J_{o_m}(x_o)^T]^T$.

Under the rigid load assumption, the forces of the end-effectors projected to frame CM can be decomposed as

$$h_o = h_{os} + h_{om}, \tag{7.6}$$

where $h_o = J_{oq}^T h$ and the projection matrix $J_{oq}(x_o)$ is given by:

$$J_{oq}(x_o) = \begin{bmatrix} J_{o_1}(x_o) & \cdots & 0 \\ \vdots & \ddots & \vdots \\ 0 & \cdots & J_{o_m}(x_o) \end{bmatrix}.$$

In Eq. 7.6, $h_{os} \in X_s$ is the vector of squeeze forces [12] and $h_{om} \in X_m$ is the vector of motion forces. The subspace X_m ($\dim(X_m) = k$) is called motion subspace and its orthogonal complement, X_s ($\dim(X_m) = k(m-1)$), is called squeeze subspace. If h_o belongs to the squeeze subspace X_S, the resulting force has no contribution to the motion of the load. The squeeze subspace is given by the kernel of $W^T = [I_k\, I_k \ldots I_k] \in \Re^{k\times(km)}$, that is, $X_s = \{h_{os} \in \Re^{km} | W^T h_{os} = 0\}$. W^T transforms the mk-dimensional vector h_o into the k-dimensional vector of the resulting force at the load frame of reference CM,

$$h_{or} = W^T h_o = W^T J_{oq}^T(x_o) h.$$

Joint torques of the form $D(q)^T h_{sc}$, where $D(q) = [D_1(q_1) \ldots D_m(q_m)]$ and $h_{sc} \in X_s$, do not affect the motion of the load if the manipulators' configurations are not singular. The motion of the manipulators, however, affects the squeeze forces due to the squeeze components of the d'Alembert (inertial) forces. Therefore, the vector of squeeze forces can be further decomposed as:

$$h_{os} = h_{sm} + h_{sc}, \tag{7.7}$$

where h_{sm} is the component of the squeeze forces induced by the motion and h_{sc} is the component not affected by the motion.

In [14], a motion and force control is designed for a fully-actuated cooperative system. In that work, first the squeeze forces are ignored and a motion control with compensation of the gravitational torques is designed. Then, a squeeze force control is designed considering the component of the squeeze forces caused by the motion (h_{sm}) as disturbances. In the next section, we develop a similar control strategy for cooperative manipulators with passive joints.

7.3 Control of Cooperative Arms with Passive Joints

In the presence of passive joints, the cooperative manipulator control problem can be decomposed in two components, namely, motion control and squeeze force control. A stable motion control with compensation of the gravitational torques is designed first. For this purpose, the Jacobian matrix $Q(q)$ that relates the velocities in the active joints to the load velocities (or equivalently, from the virtual work principle, the torques in the actuated joints to the resulting forces in the load), has to be calculated. The control law applied to the actuated joints is given by

$$\tau_a = \tau_{mg} + \tau_s, \tag{7.8}$$

where τ_a is the vector of torques at the active joints, τ_{mg} is the motion control law with compensation for the gravitational torques, and τ_s is the squeeze force control law.

In the following, a method to compute the Jacobian matrix $Q(q)$ for cooperative robotic systems with $m > 1$ is shown. From Eq. 7.2,

$$m\dot{x}_o = D_1(q_1)\dot{q}_1 + D_2(q_2)\dot{q}_2 + \cdots + D_m(q_m)\dot{q}_m. \tag{7.9}$$

Assume that, among all n joints in all manipulators, n_a joints are actuated and n_p joints are passive. The positions of the passive joints are grouped in the vector q_p and the positions of the actuated joints are grouped in the vector q_a (see Fig. 7.1, where q_{ij} represents the jth joint of the manipulator i). Partitioning Eq. 7.9 in quantities related to the passive and actuated joints, we obtain:

$$m\dot{x}_o = \sum_{i=1}^{m} D_{ai}(q_i)\dot{q}_a + \sum_{i=1}^{m} D_{pi}(q_i)\dot{q}_p = D_a(q)\dot{q}_a + D_p(q)\dot{q}_p, \tag{7.10}$$

where a refers to the actuated joints and p to the passive joints. Considering again Eq. 7.2, we can find two more expressions relating the velocities of the passive and active joints. When m is even,

$$\sum_{i=1}^{m}(-1)^{i+1}D_i(q_i)\dot{q}_i = 0, \tag{7.11}$$

which can be partitioned as

$$\sum_{i=1}^{m}(-1)^{i+1}D_{ai}(q_i)\dot{q}_a + \sum_{i=1}^{m}(-1)^{i+1}D_{pi}(q_i)\dot{q}_p = R_a(q)\dot{q}_a + R_p(q)\dot{q}_p = 0. \tag{7.12}$$

It is interesting to note that such relationship cannot always be found in individual manipulators with passive joints [7]. When m is odd,

$$\sum_{i=1}^{m}(-1)^{i+1}D_i(q_i)\dot{q}_i = \dot{x}_o, \tag{7.13}$$

which can also be partitioned in terms of the actuated and passive joints as

$$R_a(q)\dot{q}_a + R_p(q)\dot{q}_p = \dot{x}_o. \tag{7.14}$$

Using Eqs. 7.10, 7.12, and 7.14, the velocities of the load are related to the velocities of the actuated joints by:

$$\dot{x}_o = Q(q)\dot{q}_a, \tag{7.15}$$

where

$$Q(q) = \frac{1}{m}\Big(D_a(q) - D_p(q)R_p^{\#}(q)R_a(q)\Big), \tag{7.16}$$

if m is even, and

$$Q(q) = \Big(mI - D_p(q)R_p^{\#}(q)\Big)^{-1}\Big(D_a(q) - D_p(q)R_p^{\#}(q)R_a(q)\Big), \tag{7.17}$$

if m is odd. $R_p^{\#}(q)$ represents the pseudo-inverse of $R_p(q)$.

One can observe that $R_p(q)$ is of dimension $k \times n_p$ where $k \geq n_p$. For the pseudo-inverse of R_p to exist the matrix must be full rank. Therefore, in this chapter we make the following assumptions:

Assumption 1 The number of actuated joints n_a is not smaller than the dimension of the load coordinates k, which in turn is not smaller than the number of passive joints n_p, i.e., $n_a \geq k \geq n_p$.
Assumption 2 The matrix $R_p(q)$ is full rank.

Assumption 2 means that singularities in $R_p(q)$, which are determined by the number and position of the passive joints, must be avoided. The robot configurations where $R_p(q)$ is not full rank are discussed in [7]. Examples with cooperative systems formed by two planar manipulators and by two Puma robots show that the regions where $R_p(q)$ is not full rank are limited.

The Jacobian matrix $Q(q)$ has important kinematic properties and can be used to study the manipulability of cooperative system with passive joints [1].

7.3.1 Motion Control

In this section, we present a stable motion control method that includes compensation of the gravitational torques, but ignores the effects of the squeeze forces on the load. The matrix $Q(q)$ is used to design motion forces proportional to the load position and velocity errors in the actuated joints space. The gravity forces and torques in the load and in the passive joints are compensated by the actuated joints using the Jacobian matrices $R_p(q)$, $R_a(q)$, and $Q(q)$. The motion control law for the actuated joints is given by:

$$\tau_{mg} = \tau_M + \tau_g, \tag{7.18}$$

with the motion component given by

$$\tau_M = Q^T(K_p\Delta x_o + K_v\Delta\dot{x}_o), \tag{7.19}$$

where $\Delta x_o = (x_o^d - x_o)$ is the load position error, x_o^d is the desired position of the load, the diagonal matrices K_p and K_v are positive definite, $\Delta\dot{x}_o = (\dot{x}_o^d - \dot{x}_o)$ is the load velocity error, and $\dot{x}_o^d$ is the desired velocity of the load.

The term that compensates for the gravitational torques is given by:

$$\tau_g = g_a - R_o^T g_p + (J_a^T - R_o^T J_p^T) f_o, \tag{7.20}$$

where

$$R_o = \begin{cases} R_p^{\#} R_a & \text{if } m \text{ is even,} \\ R_p^{\#}(R_a - Q) & \text{if } m \text{ is odd,} \end{cases}$$

and f_o is an mk-dimensional vector chosen to satisfy

$$J_o^T f_o = b_o. \tag{7.21}$$

The gravitational terms g_a and g_p are computed considering the following partition of the combined dynamics of the manipulators, Eq. 7.4:

$$\bar{M}\ddot{\bar{q}} + \bar{C}\dot{\bar{q}} + \bar{g} = \bar{\tau} - \bar{J}h, \tag{7.22}$$

where $\bar{q} = [q_a^T \; q_p^T]^T$, $\bar{g} = [g_a^T \; g_p^T]^T$, $\bar{J} = [J_a \; J_p]^T$, $\bar{\tau} = [\tau_a^T \; \tau_p^T]^T$, $\bar{C} = [C_a^T \; C_p^T]^T$, and

$$\bar{M} = \begin{bmatrix} M_{aa} & M_{ap} \\ M_{pa} & M_{pp} \end{bmatrix}.$$

Based on the proposed motion control law, the following theorem is stated.

Theorem *Consider that Assumptions 1 and 2 are satisfied and the desired trajectories belong to* $S = \{\dot{x}_o^d(t), \ddot{x}_o^d(t) \in L_2([0,\infty)) : \dot{x}_o^d(t)$ *and* $\ddot{x}_o^d(t)$ *are uniformly continuous* $\}$. *Let the control law be given by Eqs.* 7.18–7.20, *then*:

(a) *The cooperative system is asymptotically stable, i.e., the load velocity converges to zero as* $t \to \infty$;

(b) *The position error* Δx_o *converges to the manifold given by*

$$Q^T K_p \Delta x_o + (J_a^T - R_o^T J_p^T) J_{oq}^{-T} h_{sc} = 0.$$

Proof Consider first a class of desired trajectories with desired velocities equal to zero (set-point control problem) and the following Lyapunov function candidate:

$$V = \frac{1}{2}\dot{x}_o^T M_o \dot{x}_o + \frac{1}{2}\dot{\bar{q}}^T \bar{M} \dot{\bar{q}} + \frac{1}{2}\Delta x_o^T K_p \Delta x_o, \tag{7.23}$$

where the sum of the first two terms are equal to the kinetic energy of the system. Differentiating Eq. 7.23 and considering Eqs. 7.5 and 7.22:

$$\dot{V} = -\dot{x}_o^T b_o - \dot{q}_a^T g_a - \dot{q}_p^T g_p + \dot{q}_a^T \tau_a + \Delta x_o^T K_p \Delta \dot{x}_o, \tag{7.24}$$

where the fact that $(\dot{\bar{M}} - 2\bar{C})$ is skew-symmetric was used. Substituting (7.18) in (7.24),

$$\dot{V} = -\dot{x}_o^T K_v \dot{x}_o \leq 0 \tag{7.25}$$

which, by LaSalle's Invariance Principle, implies the asymptotic convergence of $\dot{x}_o(t)$ to zero. Therefore, the load always goes to the steady state under the control law given by (7.18).

Consider now the trajectory tracking control problem. The following Lyapunov function candidate is selected:

$$V = \frac{1}{2}\Delta\dot{x}_o^T M_o \Delta\dot{x}_o + \frac{1}{2}\Delta\dot{\bar{q}}^T \bar{M} \Delta\dot{\bar{q}} + \frac{1}{2}\Delta x_o^T K_p \Delta x_o, \tag{7.26}$$

where $\Delta\dot{\bar{q}} = (\dot{\bar{q}}^d - \dot{\bar{q}})$, $\dot{\bar{q}}^d$ is the projection of $\dot{x}_o^d$ in the joint space obtained using (7.10), and $\bar{D} = [D_a \; D_p]$. It is important to observe that the errors $\Delta\bar{q}$ and $\Delta\dot{\bar{q}}$ are not present in the control law, and $\Delta\dot{\bar{q}}$ is used here only to prove the stability of the trajectory tracking controller. Differentiating (7.26), considering (7.5) and (7.22), and applying the control law given by (7.18)–(7.20):

$$\begin{aligned}\dot{V} = &(\dot{x}_o^d)^T M_o \ddot{x}_o^d - \dot{x}_o^T M_o \ddot{x}_o^d + (\dot{\bar{q}}^d)^T \bar{M} \ddot{\bar{q}}^d - \dot{\bar{q}}^T \bar{M} \ddot{\bar{q}}^d \\ &+ (\dot{\bar{q}}^d)^T \left(\bar{C} - \frac{1}{2}\dot{\bar{M}}\right)\dot{\bar{q}} + \frac{1}{2}(\dot{\bar{q}}^d)^T \dot{\bar{M}} \dot{\bar{q}}^d - \frac{1}{2}\dot{\bar{q}}^T \dot{\bar{M}} \dot{\bar{q}}^d \\ &- (\dot{x}_o^d)^T K_v \dot{x}_o^d + 2(\dot{x}_o^d)^T K_v \dot{x}_o - \dot{x}_o^T K_v \dot{x}_o.\end{aligned} \tag{7.27}$$

Since K_v is symmetric positive definite, $\dot{x}_o^T K_v \dot{x}_o$ and $(\dot{x}_o^d)^T K_v \dot{x}_o^d$ satisfy the following inequalities:

$$\dot{x}_o^T K_v \dot{x}_o \geq k_v \|\dot{x}_o\|, \tag{7.28}$$

$$(\dot{x}_o^d)^T K_v \dot{x}_o^d \geq k_v \|\dot{x}_o^d\|, \tag{7.29}$$

where k_v is the smallest eigenvalue of K_v. Therefore, at any given at instant t,

$$\dot{V}(t) \leq \dot{x}_o^d(t)^T \vartheta_1(t) + \dot{x}_o(t)^T \vartheta_2(t) - k_v \|\dot{x}_o^d(t)\|^2 - k_v \|\dot{x}_o(t)\|^2, \tag{7.30}$$

where $\vartheta_1(t)$, and $\vartheta_2(t)$ are terms dependent on the model parameters and the desired trajectory. If the desired trajectories belong to S, $\vartheta_1(t)$ and $\vartheta_2(t) \in L_2([0,\infty))$. Integrating both sides of (7.30) from t_0 to t, and considering that the inner product satisfies the Cauchy-Schwarz inequality:

$$\begin{aligned} V(t) - V(t_0) \leq & \|\vartheta_1(t)\|_{L_2([t_0,t])} \|\dot{x}_o^d\|_{L_2([t_0,t])} - k_v \|\dot{x}_o^d\|^2_{L_2([t_0,t])} - k_v \|\dot{x}_o\|^2_{L_2([t_0,t])} \\ & + \|\vartheta_2(t)\|_{L_2([t_0,t])} \|\dot{x}_o\|_{L_2([t_0,t])}. \end{aligned} \tag{7.31}$$

Completing the squares in the above equation:

$$\begin{aligned} V(t) - V(t_0) \leq & -k_v \left(\|\dot{x}_o^d\|_{L_2([t_0,t])} - \frac{\|\vartheta_1(t)\|_{L_2([t_0,t])}}{2k_v} \right)^2 \\ & - k_v \left(\|\dot{x}_o\|_{L_2([t_0,t])} - \frac{\|\vartheta_2(t)\|_{L_2([t_0,t])}}{2k_v} \right)^2 \\ & + \frac{\|\vartheta_1(t)\|^2_{L_2([t_0,t])}}{4k_v} + \frac{\|\vartheta_2(t)\|^2_{L_2([t_0,t])}}{4k_v}. \end{aligned} \tag{7.32}$$

From this equation, we see that $V(t)$ is bounded by $V(t_0)$ plus the third and forth terms on the right side of (7.32). Since $V(t) \geq 0$ (7.26), $\vartheta_1(t) \in L_2([0,\infty))$, and $\vartheta_2(t) \in L_2([0,\infty))$, then (7.32) implies that $V(t)$ is uniformly bounded for all $t > 0$, which implies that Δx_o, $\Delta \dot{x}_o$, $\Delta \bar{q}$ are all uniformly bounded. Still from (7.32), if $V(t)$ is uniformly bounded, then $\dot{x}_o \in L_2([0,\infty))$, $\dot{x}_o$ is uniformly continuous, and $\dot{x}_o$ is convergent to zero as $t \rightarrow \infty$. Thus, the load always converges to the desired trajectories that belong to S, under the control law given by (7.18). Consider now part (b) of the theorem. Substituting the second line of (7.22) in the first line, for $\dot{q} = \ddot{q} = 0$, one obtains:

$$Q^T K_p \Delta x_o + (J_a^T - R_o^T J_p^T) h + (J_a^T - R_o^T J_p^T) f_o = 0. \tag{7.33}$$

Substituting (7.21) in (7.5), for $\dot{x}_o = \ddot{x}_o = 0$, and selecting f_o that results in a null squeeze at the frame CM:

$$f_o = -J_{oq}^{-T} h_{om}. \tag{7.34}$$

Finally, substituting (7.34), (7.6), and (7.7) in (7.33), one obtains:

$$Q^T K_p \Delta x_o + (J_a^T - R_o^T J_p^T) J_{oq}^{-T} h_{sc} = 0. \tag{7.35}$$

□

7.3.2 Squeeze Force Control

In this section, the squeeze force control is designed considering the component of the squeeze forces generated by the motion, h_{sm}, as disturbances. This assumption is realistic since h_{sm} is not affected by joint torques of the form $D^T h_{sc}$ Eq. 7.7.

Here, since h_{sc} is an mk–dimensional vector and $\dim(X_s) = k(m-1)$, the control law for the actuated joints can be written as:

$$\tau_s = -D_{sa}^T A_c^T \gamma_s, \tag{7.36}$$

where the image of A_c^T projects the null space of A^T, i.e., $\mathrm{Im}(A_c^T) = X_s$, the $k(m-1)$-dimensional vector γ_s is the squeeze force control variable (computed later as a function of the measured squeeze forces Γ_s), and

$$D_{sa} = \begin{bmatrix} D_{a1} & & 0 \\ & \ddots & \\ 0 & & D_{am} \end{bmatrix},$$

where D_{ai} relates velocities of the actuated joints of manipulator i to load velocities.

Note that n_p constraints are imposed by the passive joints in γ_s, i.e.,

$$0_{n_p} = -D_{sp}^T A_c^T \gamma_s, \tag{7.37}$$

where D_{sp} relates velocities of the passive joints and load velocities.

In the fully-actuated cooperative system, k inputs are needed to control the k components of the motion of the load and $k(m-1)$ inputs are utilized to control the squeeze forces.

For the cooperative system with passive joints, if the manipulators are not kinematically redundant, it is not possible to independently control all components of the squeeze forces. As n_p constraints are imposed by Eq. 7.37,

the number of components of the $k(m-1)$–dimensional vector of measured squeeze forces Γ_s that can be independently controlled is:

$$n_s = \begin{cases} k(m-1) - n_p = n_a - k & \text{if } n_a > k, \\ 0 & \text{otherwise.} \end{cases}$$

If $n_a > k$, n_s components of Γ_s can be independently controlled. In this case, the vector γ_s can be partitionated by a permutation matrix P_{sd}, i.e.,

$$P_{sd}\gamma_s = \begin{bmatrix} \gamma_{sc} \\ \gamma_{sn} \end{bmatrix}, \tag{7.38}$$

where γ_{sc} is the n_e-dimensional vector of the independently controlled components and γ_{sn} is the n_p-dimensional vector computed as a function of γ_{sc} using Eqs. 7.37 and 7.38.

To compute the vector γ_{sc} recall that, if an asymptotically stable motion control law is utilized, h_{sm} goes to zero as $t \to \infty$. As the transient performance and convergence rate of h_{os} are influenced by h_{sm} in a feedback control approach, [14] suggests a pre-processing of the measured squeeze forces Γ_s by a strictly proper linear filter, such as an integrator. Then, γ_{sc} is given at time t by:

$$\gamma_{sc}(t) = \Gamma_{scd}(t) + K_i \int_{s=t_0}^{s=t} \left(\Gamma_{scd}(s) - \Gamma_{sc}(s)\right) ds, \tag{7.39}$$

where K_i is a positive diagonal matrix, $\Gamma_{sc}(s)$ is the vector formed by the independently controlled components of $\Gamma_s(s)$, and $\Gamma_{scd}(s)$ is the vector of their desired values. The squeeze force control is given by Eq. 7.36 with γ_{sc} computed as in Eqs. 7.37–7.39.

7.4 Dynamic Load-Carrying Capacity of Cooperative Manipulators with Passive Joints

Since one of the main reasons for using more than one manipulator is the manipulation of heavy loads, it is important to ensure that a multi-arm system with passive joints can execute the task. This can be verified by calculating the dynamic load-carrying capacity (DLCC). The DLCC is defined as the maximum load that can be carried by the system over a given trajectory. When the robots lose one or more actuators, the DLCC generally decreases. Here we extend the DLCC concept, originally conceived for fully-actuated cooperative systems [15], to underactuated ones. The DLCC of the system with passive joints is calculated via the algorithm presented in [15], except that a new linear programming problem is defined taking into account torque constraints. It is important to observe that the DLCC is obtained based on the desired trajectory and known parameters of the load.

From Eq. 7.5 we can write:

$$h_{or} = \begin{bmatrix} m_o I & 0 \\ 0 & I_o \end{bmatrix} \begin{bmatrix} \ddot{p}_o \\ \dot{\omega}_o \end{bmatrix} + \begin{bmatrix} m_o g \\ \omega_o \times (I_o \omega_o) \end{bmatrix}, \tag{7.40}$$

where $h_{or} = J_o(x_o)^T h$ is the resulting force vector at the frame of reference CM attached to the load, the terms on the left side of Eq. 7.5 were expanded, I_o is the inertia matrix of the load, m_o is the mass of the load, ω_o is the vector of angular velocities of the load, and g is the gravity vector. Considering loads with inertia matrix equal to $I_o = I_{oc} m_o$, where I_{oc} is constant, (7.40) can be written as:

$$h_{or} = \begin{bmatrix} g + \ddot{p}_o \\ I_{oc}\dot{\omega}_o + \omega_o \times (I_{oc}\omega_o) \end{bmatrix} m_o. \tag{7.41}$$

As the load has k components of motion, the first k components (partition K) of the joint space (with n joints) are chosen as generalized coordinates. Then, the dynamics of the joints in partition K is given by:

$$\tau_k + J_k^{n-k}(q)^T \tau_{n-k} + D_k^{o}(q)^T h_{or} - J_k^{n}(q)^T (M(q)\ddot{q} + C(q,\dot{q})\dot{q} + g(q)) = 0, \quad (7.42)$$

where τ_k is the vector of torques of the joints in partition K, τ_{n-k} is the vector of torques of the joints that do not belong to partition K, $J_k^{n-k}(q)$ is the Jacobian matrix that relates the velocities of the joints in partition K to the velocities of the joints that do not belong to partition K, $D_k^{o}(q)$ is the Jacobian matrix that relates the velocities of the joints in partition K to the velocities of the load, and $J_k^{n}(q)$ is the Jacobian matrix that relates the velocities of the joints in partition K to the velocities of all joints of the system.

Substituting Eq. 7.41 in 7.42, one obtains:

$$A_t \tau + a_o m_o - J_k^{n}(q)^T (M(q)\ddot{q} + C(q,\dot{q})\dot{q} + g(q)) = 0, \quad (7.43)$$

where $\tau = [\tau_k^T \; \tau_{n-k}^T]^T$, $A_t = \begin{bmatrix} I & J_k^{n-k}(q)^T \end{bmatrix}$, and

$$a_o = D_k^{o}(q)^T \begin{bmatrix} g + \ddot{p}_o \\ I_{oc}\dot{\omega}_o + \omega_o \times (I_{oc}\omega_o) \end{bmatrix}.$$

It is possible to rewrite Eq. 7.43 as:

$$Ax = b, \quad (7.44)$$

where $A = [A_t \; a_o]$, $x = [\tau^T \; m_o]^T$, and $b = J_k^{n}(q)^T (M(q)\ddot{q} + C(q,\dot{q})\dot{q} + g(q))$.

As the number of constraints in x is greater than the number of equations, Eq. 7.44 resembles a linear system with equality constraints on the load mass and joint torques. The constraints on x are:

$$m_o > 0, \quad (7.45)$$

and

$$|\tau_j| \begin{cases} \leq \tau_{max_j} & \text{if joint } j \text{ is actuated} \\ = 0 & \text{if joint } j \text{ is passive} \end{cases}, \; j = 1, \ldots, n, \quad (7.46)$$

where τ_{max_j} is the maximum torque applied at joint j. One can observe that the presence of passive joints is addressed only in the constraints given by Eq. 7.46, which is the main difference of this method to the one proposed in [15] for the fully-actuated system.

Equations 7.44–7.46 impose constraints in the linear programming problem to be solved. As the maximum load mass should be found for each desired trajectory, the linear programming problem is:

$$\max(c^T x), \quad (7.47)$$

where $c^T = [0_{n\times 1} \; 1]$.

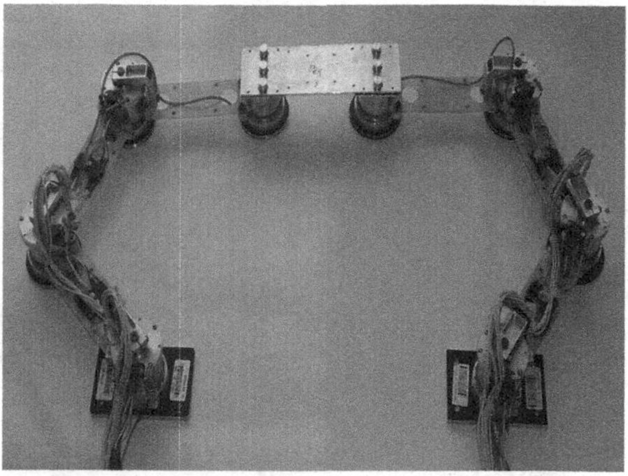

Fig. 7.2 Cooperative manipulator system composed of two UARM manipulators handling a common load

The procedure to find the DLCC over a given desired trajectory can be summarized as follows:

(a) design the desired trajectory;
(b) solve the linear programming problem for each sampling time of the desired trajectory to obtain the optimal torques and the maximum load mass, subject to the constraints imposed by (7.44–7.46);
(c) record the mass of the load obtained in each sampling time;
(d) compute the DLCC for the desired trajectory as the minimum value of the mass recorded.

7.5 Examples

In this section, the control system developed in Sect. 7.3 is applied to a planar cooperative manipulator system composed of two individual UARM manipulators (Fig. 7.2) in two different configurations, namely $n_p = 1$ and $n_p = 3$. The results of the method developed in Sect. 7.4 to analyze the DLCC are also presented.

7.5.1 Design Procedures

The Cooperative Manipulator Control Environment (CMCE), Fig. 7.3 allows the user to simulate and control the actual cooperative system using one single graphical user interface.

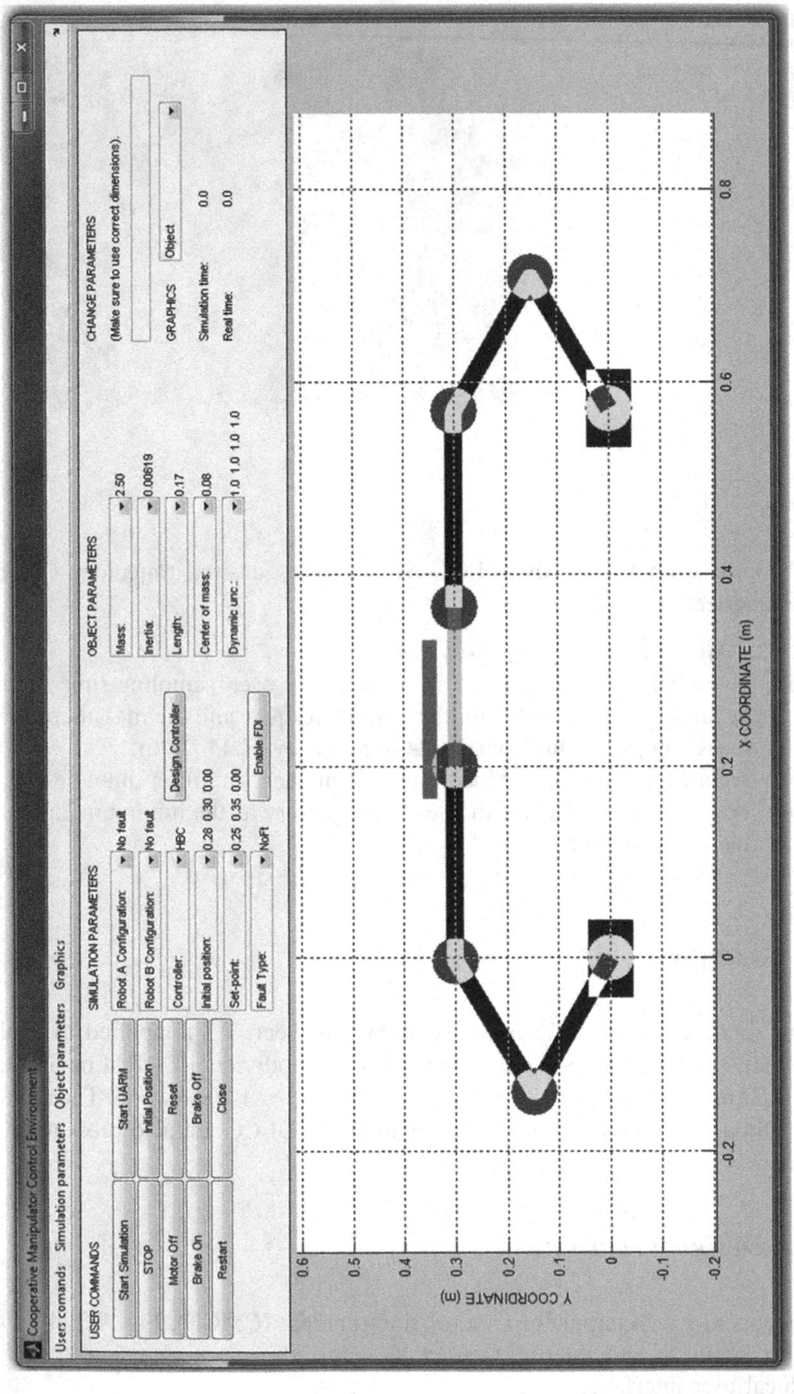

Fig. 7.3 Cooperative Manipulator Control Environment

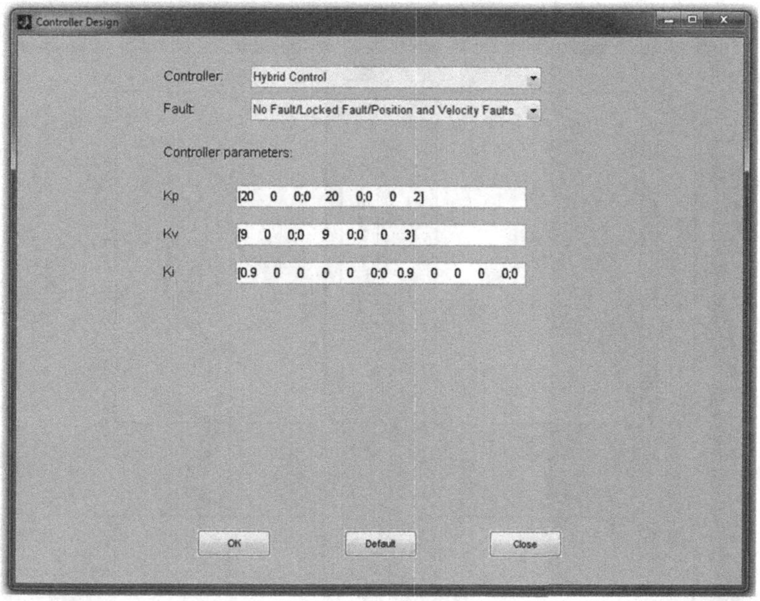

Fig. 7.4 Controller Design box for Hybrid Control of a cooperative manipulator system

In the menus *Robot A Configuration* and *Robot B Configuration*, the user can select in which joints of robots A (on the left) and B, respectively, the faults will occur. The following options are available in the menu *Fault Type*:

- *No Fault*: the cooperative system operates without faults;
- *Free-Swinging Joint*: one or more joints of the cooperative system lose actuation, and the torque applied to these joints is set to zero;
- *Locked Joint*: one or more joints of the cooperative system are locked;
- *Incorrect Position*: the joint position measurements are corrupt or missing;
- *Incorrect Velocity*: the joint velocity measurements are corrupt or missing.

The last three fault types are discussed in Chap. 8, together with the fault detection and isolation algorithm proposed for cooperative manipulator systems. Here, we are only interested in controlling cooperative manipulators with free-swinging joint, as discussed previously. In the CMCE, the control strategy presented in this chapter is named Hybrid Control (HBC) and defined as the default controller under the menu *Controller*. The controller's parameters (K_p, K_v and K_i) can be modified by clicking on the push button *Design Controller* (Fig. 7.3) and selecting the appropriate fault configuration Fig. 7.4. The other available control strategies (NLH—quasi-LPV, NLH—Game Theory and NLH—Neural Network) will be presented in Chap. 9.

To implement the Hybrid Control strategy for cooperative manipulators with passive joints, the main programming issues to be taken into account are the computation of matrix Q, Eq. 7.15, and the implementation of the squeeze force control,

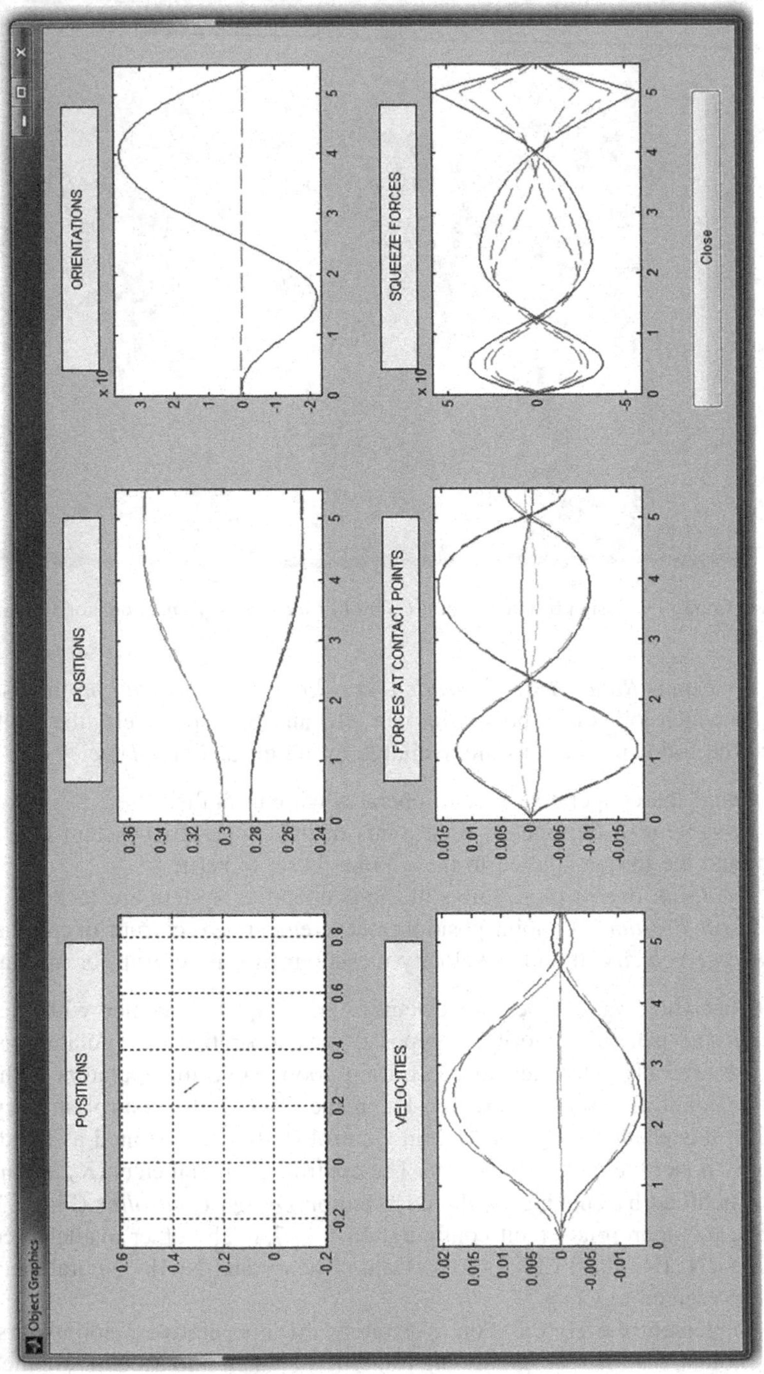

Fig. 7.5 Load position, velocity, orientation, and forces at the end of a simulation

Eq. 7.36. The following MATLAB® code, extracted from the file `uarm_cont.m` deals with both issues.

File: uarm_cont.m

```
% Compute position and force errors
error = xo_d(:,i)-xo_est(:,i);
derror = dxo_d(:,i)-dxo_est(:,i);
sum_force_error=sum_force_error + ...
    (cont.squeeze_setpoint-fs_est(:,i-1));

...


% Verify in which joint the fault is set to occur
if (fault.joint_a>0)
     passive(fault.joint_a) = 1;
end
if (fault.joint_b>0)
     passive(arm_a.n+fault.joint_b) = 1;
end

npa = 0; % Number of passive joints
nat = 0; % Number of active joints
q = [qa_est(:,i);qb_est(:,i)];
dq = [dqa_est(:,i);dqb_est(:,i)];
V = [ Va_est ; Vb_est];
D = [Da_est , -Db_est];
R = [Da_est , Db_est];
J = [Ja_est  zeros(arm_a.n,obj.k); ...
    zeros(arm_b.n,obj.k) Jb_est ];
dJ = [-dJa_est zeros(n/2); zeros(n/2) -dJb_est];

% Perform matrix partition
for i_aux=1:(arm_a.n+arm_b.n),
    if passive(i_aux)==1, % Passive joint
        Dpa(:,npa+1) = D(:,i_aux);
        Rpa(:,npa+1) = R(:,i_aux);
        Jpa(:,npa+1) = J(:,i_aux);
        dJpa(:,npa+1) = dJ(:,i_aux);

        if npa==0
            qpa = q(i_aux);
            dqpa = dq(i_aux);
        else
            qpa = [ qpa ; q(i_aux) ];
```

```
            dqpa = [ dqpa ; dq(i_aux) ];
        end
        npa=npa+1;
    else % Active joint
        Dat(:,nat+1) = D(:,i_aux);
        Rat(:,nat+1) = R(:,i_aux);
        Jat(:,nat+1) = J(:,i_aux);
        dJat(:,nat+1) = dJ(:,i_aux);

        if nat == 0
            qat = q(i_aux);
            dqat = dq(i_aux);;
        else
            qat = [qat; q(i_aux)];
            dqat = [dqat; dq(i_aux)];
        end
        if i_aux < arm_a.n+1, % Robot A
            if nat==0
                Dat_m = [R(:,i_aux); zeros(3,1)];
            else
                Dat_m = [Dat_m , [R(:,i_aux); ...
                 zeros(3,1)]];
            end
        else % Robot B
            if nat==0
                Dat_m = [zeros(3,1); R(:,i_aux)];
            else
                Dat_m = [Dat_m, [zeros(3,1); ...
                 R(:,i_aux)]];
            end
        end
        nat = nat+1;
    end
end

% Compute of matrix Q
Dpa_at = (pinv(Dpa))*Dat;
Q = 1/2*(Rat-Rpa*Dpa_at);

...

% Squeeze Force Control
Da_force = -Da_est';
```

```
Db_force = -Db_est';
fs_d=(cont.squeeze_setpoint+cont.Ki*sum_force_error);
if (fault.joint_a > 0) % Fault in Robot A

    ...

    fs_d(cont.squeze_comp_a) = ...
      -(1/(Da_force(fault.joint_a, ...
       cont.squeze_comp_a)))* ...
      (Da_force(fault.joint_a,:)*fs_d(1:arm_a.n)- ...
      Da_force(fault.joint_a,cont.squeze_comp_a)* ...
      fs_d(cont.squeze_comp_a) );
    fs_d(arm_a.n+cont.squeze_comp_a) = ...
        - fs_d(cont.squeze_comp_a);
end
if (fault.joint_b > 0) % Fault in Robot B

    ...

    fs_d(arm_a.n+cont.squeze_comp_b) =
      -(1/(Db_force(fault.joint_b, ...
       cont.squeze_comp_b)))* ...
      (Db_force(fault.joint_b,:)* ...
      fs_d(arm_a.n+1:arm_a.n+arm_b.n)- ...
      Db_force(fault.joint_b,cont.squeze_comp_b)* ...
      fs_d(arm_a.n+cont.squeze_comp_b) );
    fs_d(cont.squeze_comp_b) = ...
        - fs_d(arm_a.n+cont.squeze_comp_b);
end

% Hybrid Controller
if (all(cont.type == 'HBC'))

    % Motion Control
    tau_m = Q'*(cont.Kp*error+cont.Kv*derror);

    % Squeeze Force Control
    tau_s = [-Dat_m']*fs_d;

    tau_at = tau_m+tau_g+tau_s;
    tau_pa = zeros(npa,1);

elseif ...
```

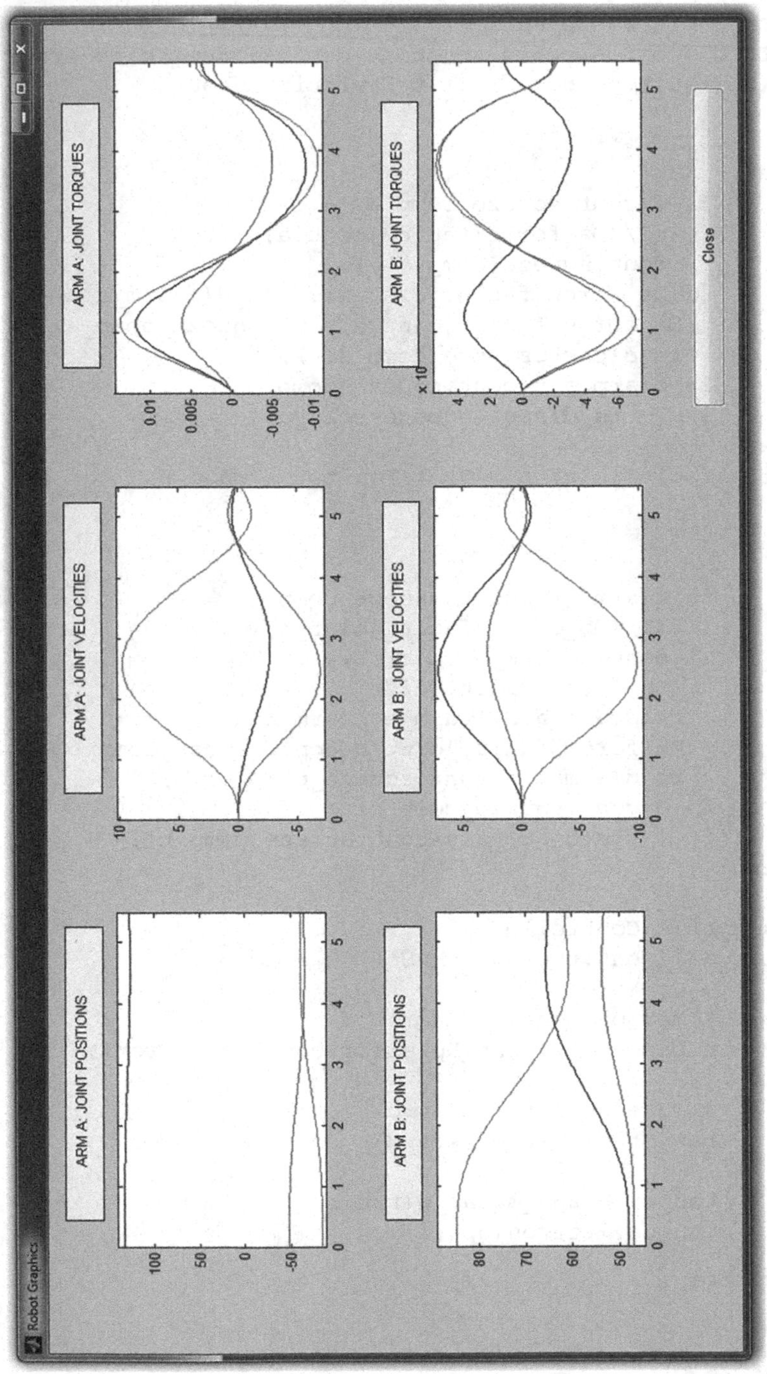

Fig. 7.6 Manipulators' position, velocity, and torques at the end of a simulation

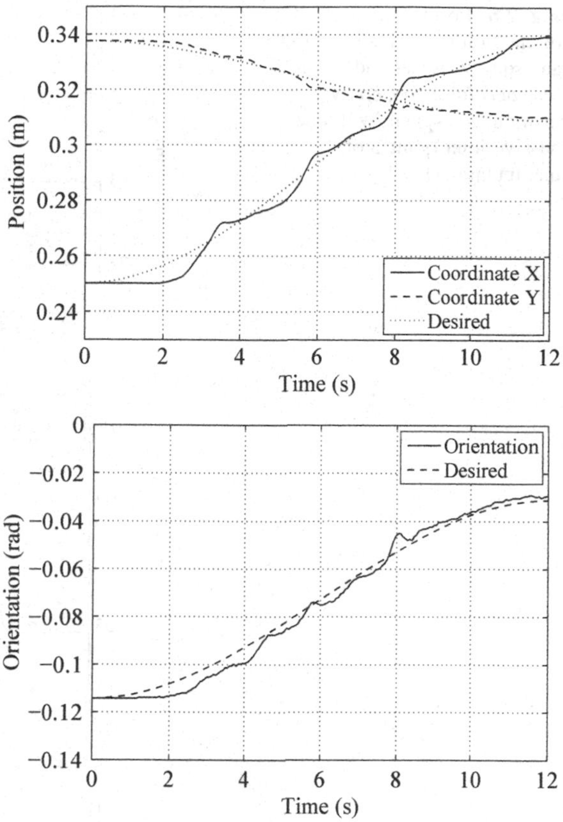

Fig. 7.7 Positions and orientation of the load for a cooperative manipulator system with three passive joints (joints 2 and 3 of manipulator 1 and joint 1 of manipulator 2)

The results obtained after performing a given simulation can be analyzed trough two graphics generated by the control environment. Activating the menu GRAPHICS (on the right of the graphical interface) and selecting the options *Object* or *Robots*, the respective graphic is created. The Object Graphics, Fig. 7.5 presents the Cartesian position, velocity, and orientation of the load's frame of reference as a function of time. The forces acting on the contact points and the squeeze forces are also presented. The reader will note that although forces on the contact points are not symmetrical (a resultant force is necessary to move the load), the squeeze forces are. (In reality, one controls only the components of the squeeze force acting in one direction; the opposite squeeze force is generated automatically according to Newton's third law). The Robots Graphics, Fig. 7.6, presents the joint position, velocities and applied torques for robots A and B. If the user has performed a free-swinging joint simulation, a zero torque value is observed in the applied torque graphic for the passive joint.

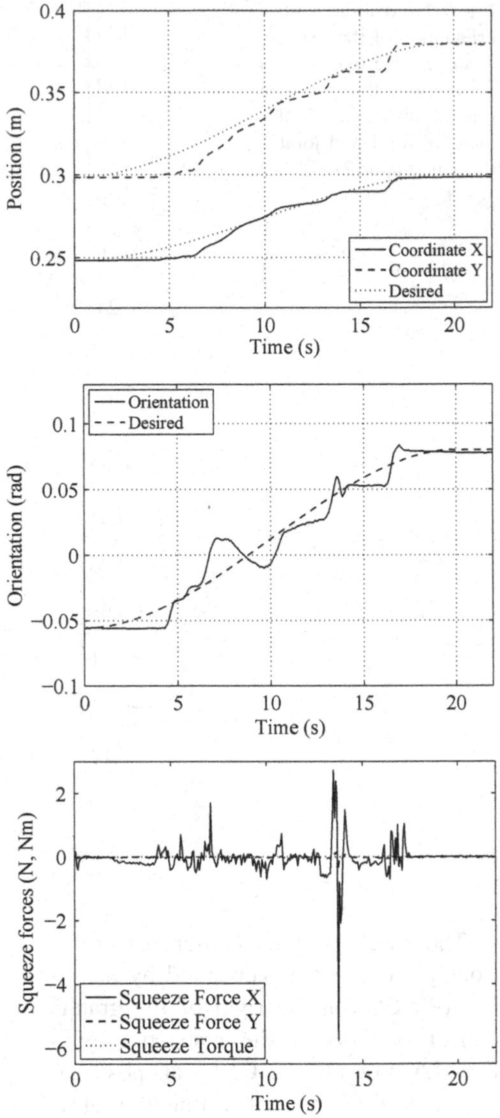

Fig. 7.8 Positions and orientation of the load (*top*) and squeeze forces and torque (*bottom*) for a cooperative manipulator system with one passive joint (joint 2 of manipulator 1)

7.5.2 Motion and Squeeze Force Control

The control methodology developed in this chapter was applied to the cooperative system composed by two UARM manipulators shown in Fig. 7.2. The first test was made with three passive joints, namely, joints 2 and 3 of manipulator 1 and joint 1 of manipulator 2. The goal is to show that even when half of all joints are passive, the position of the load can be controlled to a desired set-point. The load is a thin

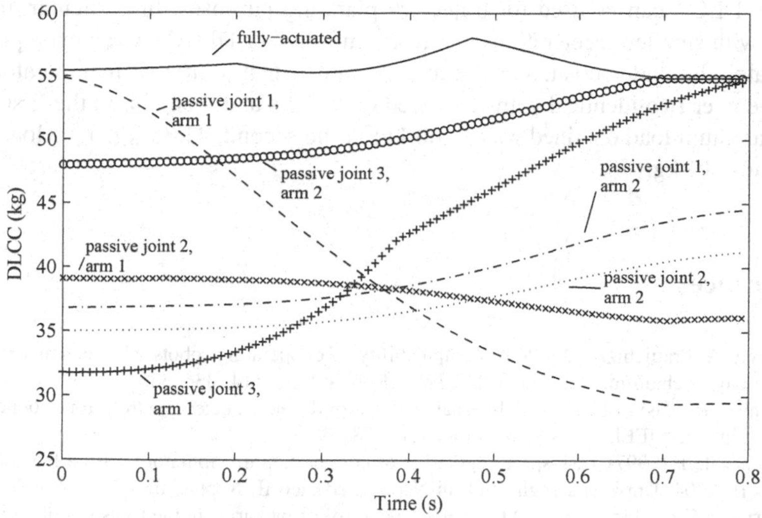

Fig. 7.9 DLCC for a fully-actuated cooperative manipulator systems and for six different underactuated configurations

disc with a diameter of 12 cm and weighing 25 g. With three active joints available to control the load's position and orientation, the squeeze forces cannot be controlled. Figure 7.7 shows an example where the positions and orientation of the load follow a desired trajectory.

In the second test only joint 2 of manipulator 1 was considered passive. With a total of five active joints, here we are able to control not only the load's position and orientation but also two components of the squeeze force. The load is a 4-component dynamometer (Kistler model 9272) with a mass of 4.2 kg. Figure 7.8 shows the resulting positions and orientation of the load and the squeeze forces. One can observe that even with this relatively heavy load, the positions and orientation are correctly controlled. As there is one passive joint, one component of the squeeze force is not controlled. In this case, the uncontrolled force in the x-axis presents greater variation when compared to the controlled force in the y-axis and the squeeze torque.

7.5.3 DLCC of Two 3-Joint Manipulators with Passive Joints

In this section, the method presented in Sect. 7.4 is used to analyze the DLCC of the underactuated cooperative system along a desired trajectory. The maximum torque in each actuated joint is equal to 25 Nm. Figure 7.9 shows simulated results of the DLCC in seven different configurations: the first with the fully-actuated system and the others with one passive joint (one simulation for each joint). One can observe that the DLCC is different for each case, indicating how much each joint contributes to the "total" DLCC.

The DLCC can be used for trajectory planning purposes. In particular, trajectories with smaller accelerations can result in a higher DLCC. As an example, we simulated the underactuated cooperative system when joint 1 of manipulator 1 is passive over two identical paths followed over 0.3 s and then 3 s. In the first case, the maximum load obtained was 31.60 kg; in the second, 41.35 kg, a payload gain of almost 10 kg.

References

1. Bicchi A, Prattichizzo D (2000) Manipulability of cooperating robots with unactuated joints and closed-chain mechanisms. IEEE Trans Rob Autom 16(4):336–345
2. Bonitz R, Hsia T (1996) Internal force-based impedance control for cooperating manipulators. IEEE Trans Rob Autom 12(1):78–89
3. Caccavale F (1997) Task-space regulation of cooperative manipulators prisma lab technical report 97-04. Università degli Studi di Napoli Federico II, Naples, Italy
4. Carignan CR, Akin DL (1988) Cooperative control of two arms in the transport of an inertial load in zero gravity. IEEE Trans Rob Autom 4(4):414–419
5. Hirano G, Yamamoto M, Mohri A (2002) Study on cooperative multiple manipulators with passive joints. In: Proceedings of the 2002 IEEE/RSJ international conference on intelligent robot and systems, Lausanne, Switzerland, pp 2855–2860
6. Koivo AJ, Unseren MA (1991) Reduced order model and decoupled control architecture for two manipulators holding a rigid object. J Dyn Syst Meas Control 113(4):646–654
7. Liu YH, Xu Y, Bergerman M (1999) Cooperation control of multiple manipulators with passive joints. IEEE Trans Rob Autom 15(2):258–267
8. Luh JYS, Zheng YF (1987) Constrained relations between two coordinated industrial robots for robot motion control. Int J Rob Res 6(3):60–70
9. Nahon M, Angeles J (1992) Minimization of power losses in cooperating manipulators. J Dyn Syst Meas Control 114(2):213–219
10. Uchiyama M (1998) Multirobots and cooperative systems. In: Siciliano B, Valavanis KP (eds) Control problems in robotics and automation. Springer-Verlag, London
11. Vukobratovic M, Tuneski A (1998) Mathematical model of multiple manipulators: cooperative compliant manipulation on dynamical environments. Mech Mach Theory 33(8):1211–1239
12. Walker ID, Freeman RA, Marcus SI (1991) Analysis of motion and internal loading of objects grasped by multiple cooperating manipulators. Int J Rob Res 10(4):396–409
13. Wang LT, Kuo MJ (1994) Dynamic load-carrying capacity and inverse dynamics of multiple cooperating robotic manipulators. IEEE Trans Rob Autom 10(1):71–77
14. Wen T, Kreutz-Delgado K (1992) Motion and force control for multiple robotic manipulators. Automatica 28(4):729–743
15. Zhao YS, Lu L, Gao F, Huang Z, Zhao TS (1998) The novel approaches for computing the dynamic load-carrying capacity of multiple co-operating robotic manipulators. Mech Mach Theory 34(4):637–643

Chapter 8
A Fault Tolerance Framework for Cooperative Robotic Manipulators

8.1 Introduction

Robotic manipulators have been deployed in an ever growing number of unstructured and/or hazardous environments, such as in outer space and in deep sea. Robots are used in these environments to limit or eliminate the presence of human beings in such dangerous places, or due to their capability to execute repetitive tasks very reliably. Faults, however, can put at risk the robots, their task, the working environment, and any humans present there. Because of the dynamic coupling between joints, inertia, and gravitational torques, faulty arms can quickly accelerate into wild motions that can cause serious damage [19]. If the controller is not designed to detect and isolate the faults, internal forces can increase and cause damage to the load or make the system unstable.

Faults occur in robots because of their inherent complexity. There are several sources of faults in robots, including electrical, mechanical, hydraulic, and software [19]. In fact, the mean-time-to-failure of industrial robots is relatively small compared to their intended life expectancy and cost. Studies in 1990s indicate that the recorded mean-time-to-failure of industrial robots was on the order of thousands of hours [3]. This number is probably smaller in unstructured or hazardous environments due to external factors, such as extreme temperatures, moving obstacles, and radiation.

In most environments, a robot can be repaired after a fault is detected. In some cases, however—say in space or under the sea—humans cannot be sent to make the necessary repairs; therefore the robot must be endowed with a minimum of fault tolerance if it is to operate robustly. This is the role of fault detection and isolation (FDI) systems—to detect faults in real-time, isolate them, and reconfigure the system so that the task (or a subset of it) can be completed despite the fault.

Probably the most obvious way to provide fault tolerance to a robotic system is physical redundancy—extra robots per task, extra joints per robot, extra motors per joint, extra computers per system, etc. These extra components not only allow the robot to continue running when any one of them fail, they also provide

A. A. G. Siqueira et al., *Robust Control of Robots*,
DOI: 10.1007/978-0-85729-898-0_8, © Springer-Verlag London Limited 2011

functionalities not available in systems with no physical redundancy. Examples include teams of two or more robots, which are capable of manipulating loads that a single robot cannot handle [21]; and manipulators with more joints than the number of degrees-of-freedom of the task, which can execute the same task in a variety of different ways—for example, minimizing energy consumption or deviating from obstacles.

It is not always possible, however, to add physical redundancy to a new robot or to retrofit existing robots so they are more fault tolerant—e.g., because of cost or size constraints, but it is generally possible to implement software-based FDI capabilities. This has been proposed by several researchers for single robotic arms (e.g. [4, 5, 11, 12, 18, 20]), cooperative manipulators [9, 13, 17] and parallel manipulators [6, 7]. A step ahead, where the FDI system is combined with a control reconfiguration scheme, was proposed for single manipulators in [4, 20]. Here we extend these results to present a fault tolerance framework that combines FDI and control reconfiguration for cooperative manipulators.

The chapter is organized as follows: Sect. 8.2 describes the kinematic and dynamic models of cooperative manipulators. Section 8.3 describes the fault tolerance framework. Section 8.4 presents the fault detection and isolation system. Section 8.5 presents the post-fault control strategies. Section 8.6 presents design procedures and results of the proposed fault tolerance framework applied to a cooperative manipulator system composed of two UARM's handling a common load.

8.2 Cooperative Manipulators

Recall from Chap. 7 that the dynamic model of a cooperative manipulator system with m arms rigidly connected to an undeformable load is given by:

$$M(q)\ddot{q} + C(q,\dot{q})\dot{q} + g(q) = \tau - J(q)^T h, \tag{8.1}$$

and the equation of motion of the load is

$$M_o\ddot{x}_o + b_o(x_o,\dot{x}_o) = J_o(x_o)^T h. \tag{8.2}$$

The cooperative system presents the following kinematic constraints:

$$x_o = \phi_i(q_i), \quad i = 1,2,\ldots,m, \tag{8.3}$$

and the n-dimensional vector of velocities of the load, $\dot{x}_o$, is constrained by

$$\dot{x}_o = D_i(x_o,q_i)\dot{q}_i, \quad i = 1,2,\ldots,m. \tag{8.4}$$

Control of cooperative manipulators is complicated by the interaction among the arms caused by force constraints. Motion of the various arms should be coordinated and the squeeze forces at the load should be controlled to avoid damages.

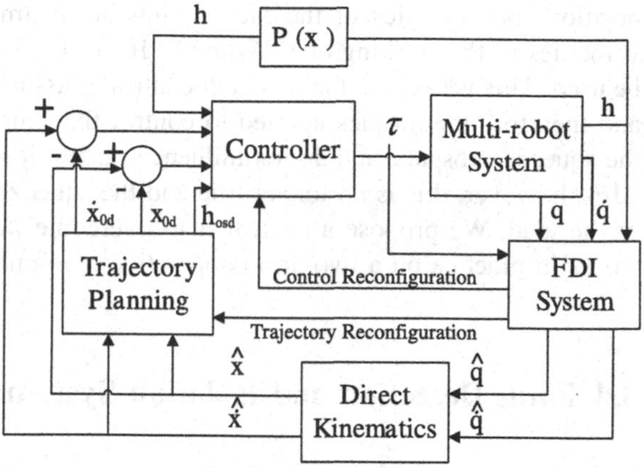

Fig. 8.1 Fault tolerance system. q_m is the vector of measured joint positions ˆ indicates an estimated quantity, the subscript d indicates desired values, and the matrix $P_s(x_o)$ converts the forces in the end-effectors to squeeze forces

Several solutions were reported in the literature to deal with the control problem of fault-free cooperative manipulators rigidly connected to an undeformable load, e. g. [1, 2, 10, 22]. When both force and motion must be controlled, hybrid control can be employed [8]. The hybrid control method for fault-free cooperative manipulators developed in [22] is particularly interesting because motion and squeeze control are independently dealt with, and because it does not utilize the inertia matrix of the robots in the control law, reducing the effect of modeling errors.

8.3 Fault Tolerance System

The block diagram of the fault tolerance system is shown in Fig. 8.1. Faults are detected and isolated by an FDI system, upon which the arms can either be locked by brakes and the trajectory re-planned starting with zero velocities; or the trajectory is re-planned without applying the brakes, starting with the current load velocity. The choice depends on the joint configurations, joint velocities, and the parameters of the cooperative system, such as maximum torque allowed and joint limits. In this chapter we show results using both options.

The fault tolerance system addresses the following categories of faults: free-swinging joint faults (FSJF), where one or more joints lose actuation and become free-swinging; locked joint faults (LJF), where one or more joints become locked; incorrectly-measured joint position faults (JPF), where the measurement of one or more joint positions is incorrect; and incorrectly-measured joint velocity faults (JVF), where the measurement of one or more joint velocities is incorrect. FSJFs and LJFs can be a result of actuator failures; and JPFs and JVFs can occur due to sensor failures.

In the framework presented here, we use the hybrid controller from [22] to control the cooperative system before the fault. When JPFs or JVFs occur, the post-fault controller is the same as the pre-fault one; the difference is that the positions or velocities of the faulty joints are estimated from the positions and velocities of the working ones. When FSJFs or LJFs occur, a new controller must be used. This is because the pre-fault controller assumes that all joints are actuated and therefore the torques applied to control the squeeze forces project only onto the squeeze subspace and do not influence the motion of the load. With FSJFs and LJFs, however, this is no longer true and the squeeze forces may produce motion of the load. We propose a controller to overcome this difficulty and show that it works in practice on a two-arm cooperative manipulator system.

8.4 Fault Detection and Isolation System

In this section, a three-step FDI system is proposed. First, JPFs are detected by analyzing the position constraints (8.3). Then, JVFs are detected by analyzing the velocity constraints (8.4). The last step is the detection of FSJFs and LJFs via two artificial neural networks (ANNs). This sequence is important because undetected JPFs can cause false detection of other faults as joint position readings are used to compute the velocities of the load in (8.4) and as inputs to the first ANN. The same occurs for undetected JVFs, as joint velocity readings are also used as inputs to the first ANN. This sequence is also important because joint velocities are generally reconstructed from joint position measurements. At each sample time, the FDI system indicates if the robots are working normally or if a fault occurred. While our framework allows for the coexistence of multiple faults of different types, we assume that faults occur one at a time.

8.4.1 Incorrectly-Measured Joint Position and Joint Velocity Faults

The load position x_o can be computed using the joint positions of an individual manipulator using Eq. 8.3. Therefore, for systems with $m > 2$ manipulators it is possible to identify the arm f with wrong joint position readings as the one whose estimate of x_o is different from the estimates of the other $m - 1$ ones. Therefore, a JPF is detected at arm f if:

$$\|\hat{x}_{of}(q_{m_f}) - \hat{x}_{oi}(q_{m_i})\| > \gamma_{p1} \quad \text{for all } i = 1, \ldots, m \text{ and } i \neq f, \tag{8.5}$$

where $\hat{x}_{oi}$ is the estimate of x_o calculated by the forward kinematics of manipulator i, q_{m_i} is the vector of measured joint positions of manipulator arm i, $\|.\|$ represents the Euclidean norm, and γ_{p1} is a positive real constant. The next step is to estimate the position of each joint $j = 1, \ldots, n_f$ of manipulator f:

$$\hat{q}_{fj} = \psi_j(q_{m_f}, \hat{x}_o), \tag{8.6}$$

where

$$\hat{x}_o = \frac{1}{m-1} \sum_{i=1, i \neq f}^{m} \hat{x}_{oi}(q_{m_i}),$$

and ψ_j is a kinematic function used to estimate the position of joint j. In a planar system with revolute joints only, ψ_j can be written as the difference between the orientation of the load and the sum of the measured positions of the joints $k \neq j$ of manipulator f. In a three-dimensional space, an inverse kinematics method can be employed to estimate $\hat{q}_{fj}$. In this case, the measured values of the joints $i \neq j$ can be used to eliminate the possible redundancy of the solutions.

Computing again the estimate of vector x_o for manipulator f for each new estimate $\hat{q}_{fj}$, the JPF at joint j of manipulator f is isolated when:

$$\|\hat{x}_o - \hat{x}_{of}(q_{m_f}, \hat{q}_{fj})\| < \gamma_{p2}, \tag{8.7}$$

where $\hat{x}_{of}(q_{m_f}, \hat{q}_{fj})$ is the vector of positions and orientations of the load estimated from the forward kinematics of manipulator f, replacing the measured position of joint j with its estimate $\hat{q}_{fj}$ and using the measured positions of the other joints, and γ_{p2} is a positive real constant.

The selection of the thresholds γ_{p1} and γ_{p2} has a strong influence in the performance of the FDI system. If the values of γ_{p1} and γ_{p2} are too small, false alarms may occur due to the presence of noise in the joint position readings. If the thresholds are too large, JPFs can go undetected. If the distribution of the noise in the joint position readings is normal and its statistical properties are known, γ_{p1} and γ_{p2} can be computed as linear functions of the variance of the noise in the joint position readings. This way, the thresholds are proportional to the variance of the noise; larger values of variance imply in larger values of thresholds and vice-versa.

The procedure to detect and isolate JPFs when $m > 2$ can be summarized as follows:

- Compare the estimate of x_o for all manipulators (8.5).
- If all values are close (in the sense of Eq. 8.5), a JPF is not declared.
- Otherwise, compute, for all joints of the faulty arm, the estimate of the joint positions (8.6) and test (8.7).
- If the test is satisfied for joint j, declare a JPF at this joint.

If $m = 2$, the arm with the fault cannot be identified just by analyzing the estimate of x_o for each arm. In this case, the joint position estimation (8.6) should be performed for the two arms using, instead of $\hat{x}_o$, the estimate obtained using the joint positions of the other arm. The same should be done in (8.7).

As it is possible to compute the velocity of the load by using the joint velocities of any arm as in Eq. 8.4, JVFs can be detected in a similar way to JPFs.

Considering the occurrence of only one fault, for $m > 2$ a JVF at joint j of arm f is detected when:

$$\|\hat{\dot{x}}_{\mathrm{o}} - \hat{\dot{x}}_{\mathrm{o}f}(\dot{q}_{m_f}, q_{m_f}, \hat{\dot{q}}_{fj})\| < \gamma_{v2}, \tag{8.8}$$

where $\hat{\dot{x}}_{\mathrm{o}f}(\dot{q}_{m_f}, q_{m_f}, \hat{\dot{q}}_{fj})$ is the velocity of the load estimated from the forward kinematics of manipulator f replacing the measured velocity of joint j with its estimate $\hat{\dot{q}}_{fj}$ and using the measured velocities ($\dot{q}_{m_f}$) of the other joints, $\hat{\dot{x}}_{\mathrm{o}}$ is the estimate of the load velocities using the measured joint velocities of the other arms, and γ_{v2} is a threshold to prevent that faults be undetected due to the presence of noise in the joint readings. As explained above, γ_{v2} is computed as a linear function of the variance of the noise in the joint velocity readings. When $m = 2$, $\dot{x}_{\mathrm{o}}$ should be replaced by the estimated velocity obtained using the joint velocities of the other arm.

The procedure to detect and isolate JVFs when $m > 2$ can be summarized as follows:

- Compare an estimate of $\dot{x}_{\mathrm{o}}$ for all manipulators using

$$\|\hat{\dot{x}}_{\mathrm{o}f}(\dot{q}_{m_f}, q_{m_f}) - \hat{\dot{x}}_{\mathrm{o}i}(\dot{q}_{m_i}, q_{m_i})\| > \gamma_{v1}, \tag{8.9}$$

  for all $i = 1, \ldots, m$ $(i \neq f)$ and γ_{v1} is a positive real constant.
- If all values are close, a JVF is not declared.
- Otherwise compute, for all joints of the faulty manipulator, the estimate of the joint velocities as:

$$\hat{\dot{q}}_{fj} = \xi_j(\dot{q}_{m_f}, q_f, \hat{\dot{x}}_{\mathrm{o}}), \tag{8.10}$$

  where

$$\hat{\dot{x}}_{\mathrm{o}} = \frac{1}{m-1} \sum_{i=1, i \neq f}^{m} \hat{\dot{x}}_{\mathrm{o}i}(\dot{q}_{m_i}, q_{m_i}).$$

- Test (8.8). If the test is satisfied for joint j, declare a JVF at this joint.

8.4.2 Free-Swinging and Locked Joint Faults

FDI systems for single manipulators generally employ the residual generation scheme to detect faults [19]. The residual vector is generated by comparing the measured states of the manipulator with their estimates obtained by a mathematical model of the fault-free system. This method, however, does not work well in the presence of modeling errors, generating false alarms or hiding fault effects.

Robust techniques [11] and artificial intelligence techniques [14] have been used to avoid these problems. In the approach presented in [18], the off-nominal behavior due to faults is mapped using an ANN trained by a robust observer based on the robot's physical model. The main problem with FDI methods that rely on the system mathematical model is that, for some real robots, detailed modeling is difficult. To overcome this problem, we proposed in [15] a method where the mathematical model of the robot is not necessary. A multilayer perceptron (MLP) is used to map the dynamics of the arm and a radial basis function network (RBFN) classifies the residual vector. The MLP mapping is static, which is possible because the states of the system are measurable, the sample time is small, and control signals are used as MLP inputs.

In [16], the strategy proposed in [15] for FDI of single manipulators is extended to cooperative manipulators. One MLP is trained to reproduce the dynamics of all arms and the load of the fault-free cooperative system (8.1). The inputs of the MLP are the joint positions, velocities, and torques at time t. The outputs of the MLP are the estimated joint velocities of the fault-free system at time $t+\Delta t$, which are compared to the measured joint velocities at time $t+\Delta t$ to generate the residual vector. The residual vector is then classified by a RBFN to provide fault information. Although the use of only one MLP in [16] is useful, the mapping of the MLP is dependent on the load parameters, such as its mass. If the system has to manipulate a different object, the ANNs have to be retrained.

Here, the dynamic model of each manipulator is mapped by one MLP. Thus, the mapping is independent of the load parameters. If the sample period Δt is sufficiently small, the dynamics of the fault-free manipulator i (7.3) can be written as:

$$\dot{q}_i(t+\Delta t) = f(\dot{q}_i(t), q_i(t), h_i(t), \tau_i(t)), \tag{8.11}$$

where $f(.)$ is a nonlinear function vector representing the dynamics of the manipulator.

Each MLP i $(i = 1, \ldots, m)$ maps the dynamic behavior of one fault-free manipulator (8.11). The inputs to the ith MLP are the joint positions, velocities, torques, and end-effector forces of manipulator i at time t. The output vector of the ith MLP, which reproduces the joint velocities of the fault-free manipulator i at time $t+\Delta t$, can be written as:

$$\hat{\dot{q}}_i(t+\Delta t) = f(\dot{q}_i(t), q_i(t), h_i(t), \tau_i(t)) + e(\dot{q}_i(t), q_i(t), h_i(t), \tau_i(t)), \tag{8.12}$$

where $e(.)$ is the vector of mapping errors. The residual of manipulator i is defined as:

$$\hat{r}_i(t+\Delta t) = \dot{q}_i(t+\Delta t) - \hat{\dot{q}}_i(t+\Delta t). \tag{8.13}$$

From (8.11) to (8.13) it can be seen that, in the fault-free case, the residual vector of manipulator i is equal to the vector of mapping errors, which must be sufficiently small when compared to the residual vector of a faulty case, in order to allow fault detection. The residual vector $\hat{r}(t+\Delta t) = [\hat{r}_1^T(t+\Delta t), \ldots, \hat{r}_m^T(t+\Delta t)]^T$ is then classified by an RBFN to provide fault information. As the residual vector of FSJFs and LJFs occurring at the same joint can occupy the same region in the input space of the RBFN, an auxiliary input ζ that gives information about the velocity of the joints is used. As there is noise in the measurement of the joint velocities, the $i = 1, \ldots, n$ component (n is the sum of the number of joints of all arms) of ζ is defined as:

$$\zeta_i(t) = \begin{cases} 1, & \text{if } |\dot{q}_i(t)| < \delta_i, \\ 0, & \text{otherwise,} \end{cases}$$

where δ_i is a threshold that can be selected based on the measurement noise. In this chapter, the RBFN is trained with Kohonen's Self-Organizing Map [15]. The fault criteria, which is used to avoid false alarms due to misclassified individual patterns, is defined as:

$$\begin{cases} \text{fault } k = 1, & \text{if } \alpha_k = \max_{j=1}^{n_o} (\alpha_j) \text{ for } d \text{ consecutive samples,} \\ \text{fault } k = 0, & \text{otherwise,} \end{cases}$$

where α_k is the output $k = 1, \ldots, n_o - 1$ of the RBFN; output n_o corresponds to normal operation.

8.5 Control Reconfiguration

The second step in the fault tolerance system for cooperative manipulators is control reconfiguration. In this stage, the controller's parameters and structure are changed according to the nature of the fault.

8.5.1 Incorrectly-Measured Joint Velocity and Position Faults

When a JPF or JVF is isolated, the sensor readings of the faulty joint are ignored and the corresponding joint position or velocity is estimated based on the kinematic constraints. As the joint positions of the faulty arm f were already estimated by the FDI system (Sect. 8.4), the component $j = 1, \ldots, n$ of the new joint position vector is defined as:

$$\hat{q}[j] = \begin{cases} \hat{q}_j, & \text{if a JPF is declared at joint } j, \\ q_m[j], & \text{otherwise,} \end{cases}$$

where $\hat{q}_j$ is the estimate of the joint j position based on the other joint readings (8.6), and $q_m[j]$ is the measured position of joint j. Similarly, the component j of the new joint velocity vector is defined as:

$$\hat{\dot{q}}[j] = \begin{cases} \hat{\dot{q}}_j, & \text{if a JVF is declared at joint } j, \\ \dot{q}_m[j], & \text{otherwise,} \end{cases}$$

where $\hat{\dot{q}}_j$ is the estimate of the joint j velocity based on the other joint readings (8.10), and $\dot{q}_m[j]$ is the measured velocity of joint j.

8.5.2 Free-Swinging Joint Faults

After a FSJF is declared, the corresponding joint acts as a passive joint, and the controller presented in Chap. 7, based on the decomposition of the motion and squeeze forces [22], can be used. A stable motion control with compensation of the gravitational torques is firstly projected ignoring the squeeze forces when $n_a \geq k$. For this purpose, a Jacobian matrix $Q(q)$, which relates the velocities in the active joints to the load velocities, is computed. Then, if the number of actuated joints is greater than the number of coordinates of motion in the load ($n_a > k$), the squeeze force controller is designed considering the component of the squeeze forces caused by the motion as a disturbance. The control law is given by:

$$\tau_a = \tau_{mg} + \tau_s, \tag{8.14}$$

where τ_{mg} is the motion control law with compensation for the gravitational torques and τ_s is the squeeze force control law. The motion controller is given by:

$$\tau_{mg} = \tau_m + \tau_g, \tag{8.15}$$

where the motion component is given by

$$\tau_m = Q^T(K_p \Delta x_o + K_v \Delta \dot{x}_o), \tag{8.16}$$

and the gravitational, centrifugal, and Coriolis compensation component is given by

$$\tau_g = g_a - (R_p^{\#} R_a)^T g_p + (J_a^T - (R_p^{\#} R_a)^T J_p^T) f_o, \tag{8.17}$$

when m is even, and

$$\tau_g = g_a - (R_p^{\#}(R_a - Q))^T g_p + (J_a^T - (R_p^{\#}(R_a - Q))^T J_p^T) f_o, \tag{8.18}$$

when m is odd. In (8.16)–(8.18), $\Delta x_o = (x_{od} - x_o)$ is the load position error, x_{od} is the desired position of the load, K_p and K_v are positive definite diagonal matrices, $\Delta \dot{x}_o = (\dot{x}_{od} - \dot{x}_o)$ is the load velocity error, and f_o is an mk-dimensional vector selected to satisfy

$$J_o^T f_o = b_o. \tag{8.19}$$

The Jacobian matrix $Q(q)$ is given by:

$$Q(q) = \frac{1}{m}\Big(D_a(q) - D_p(q) R_p(q)^{\#} R_a(q)\Big), \tag{8.20}$$

if m is even and

$$Q(q) = \left(mI - D_p(q)R_p(q)^{\#}\right)^{-1}\left(D_a(q) - D_p(q)R_p(q)^{\#}R_a(q)\right), \tag{8.21}$$

if m is odd.

The squeeze control problem when the number of actuated joints (n_a) is greater than the number of coordinates of motion in the load (k) is given by:

$$\tau_s = -D_{sa}^T A_c^T \gamma_s, \tag{8.22}$$

where

$$D_{sa} = \begin{bmatrix} D_{a1} & & 0 \\ & \ddots & \\ 0 & & D_{am} \end{bmatrix},$$

and D_{ai} relates velocities of the actuated joints of arm i and load velocities.

The vector $A_c^T \gamma_s$ gives the squeeze forces that should be applied at the load by the squeeze forces controller when there are passive joints in the arms of the cooperative system. For the cooperative system with passive joints, if the arms are not kinematically redundant, it is not possible to independently control all components of the squeeze forces. In (8.22), the components of γ_s related to the squeeze forces that are not directly controlled are computed as a function of the components that are directly controlled. The components of the desired squeeze forces that should be applied by the other arms are then computed based on the components computed for the arm f and on the geometry of the grasping.

In summary, when an FSJF is isolated and the trajectory is reconfigured (see Sect. 8.3), the controller employed for the fault-free system is switched to a new controller defined by (8.14), (8.15), and (8.22). This new controller, which is decomposed in control of motion and control of squeeze forces, is applied only in the actuated joints. See Chap. 7 for more details.

8.5.3 Locked Joint Faults

A controller similar to the one used for FSJFs can be used to control the system with locked joints. The difference is that, in the case of LJFs, one may need to replan the joint trajectories after the fault occurs. Here we assume that the trajectory planning problem is dealt with at a higher level, and focus only on guaranteeing that the load can be manipulated to its desired position.

Because the velocities of the locked joints are equal to zero, as in the case of a cooperative manipulator with passive joints we can write the Jacobian that relates the velocities of the unlocked, actuated joints to the load velocities as:

$$\dot{x}_o = Q_l \dot{q}_a, \tag{8.23}$$

where

$$Q_l = \frac{1}{m} D_a. \tag{8.24}$$

Using the same procedure described in the last section, the control law given by (8.14) and (8.15) is utilized, where:

$$\tau_m = Q_l^T (K_p \Delta x_o + K_v \Delta \dot{x}_o), \tag{8.25}$$

$$\tau_g = g_a + Q_l^T b_o, \tag{8.26}$$

and τ_s is given by (8.22).

8.6 Examples

In this section, the fault tolerant control system is applied to a cooperative manipulator system composed of two UARMs. We show experimental results for two fault configurations: a free-swinging joint fault in joint 1 of arm 1; and an incorrectly-measured joint position fault in joint 2 of arm 1.

8.6.1 Design Procedures

The proposed framework is entirely implemented in the Cooperative Manipulator Control Environment (CMCE). To enable the fault detection and isolation algorithm for a simulation where a given fault is set to occur, the user must activate the push button *Enable FDI* in the CMCE graphical interface (Fig. 8.2). When the FDI framework is enabled the fault occurs at a time during the trajectory defined by the user. This procedure is different from the results presented in Chaps. 7 and 9, where the fault is considered to be active since the beginning of the movement. The instant when the fault occurs can be defined through the specific menu displayed on the right side of the user interface.

For the free-swinging and locked joint faults, the MLP and RBFN neural networks are trained considering only the hybrid control described in Chap. 7. If the user wants to perform the FDI algorithm using another controller, the neural networks must be retrained with a set of trajectories that cover the cooperative system's workspace. This procedure is not included in the CMCE. The following MATLAB® codes implement the FDI system for the free-swinging and locked joint faults. The neural networks files (mlp_a_r2.dll, mlp_b_r2.dll and rbf_r5.dll) were generated using C++ to make the code more time-efficient.

File: uarm_fdi.m

```
...

if ((sum(Fault_pos_fdi(:,i)+Fault_vel_fdi(:,i),1)==0)&(i>1))
    % FAULT: free-swinging and locked joint
    % Normalization
    x_mlp_a(:,i) = [qa_est(:,i-1); dqa_est(:,i-1); ...
        taua_est(:,i-1); fa_est(:,i-1); -1];
    d_mlp_a(:,i) = dqa_est(:,i);
    x_mlp_b(:,i) = [qb_est(:,i-1); dqb_est(:,i-1); ...
        taub_est(:,i-1) ; fb_est(:,i-1); -1];
    d_mlp_b(:,i) = dqb_est(:,i);
    for i_aux = 1:(size(x_mlp_a,1)-1)
      x_mlp_a(i_aux,i) = (x_mlp_a(i_aux,i)-min_xa(i_aux))/...
          (max_xa(i_aux)-min_xa(i_aux)) ;
    end
    for i_aux = 1:size(d_mlp_a,1)
      d_mlp_a(i_aux,i) = (d_mlp_a(i_aux,i)- ...
          min_xa(i_aux+arm_a.n))/(max_xa(i_aux+arm_a.n)- ...
          min_xa(i_aux+arm_a.n)) ;
    end
    for i_aux = 1:(size(x_mlp_b,1)-1)
      x_mlp_b(i_aux,i) = (x_mlp_b(i_aux,i)-min_xb(i_aux))/...
          (max_xb(i_aux)-min_xb(i_aux)) ;
    end
    for i_aux = 1:size(d_mlp_b,1)
      d_mlp_b(i_aux,i) = (d_mlp_b(i_aux,i)- ...
          min_xb(i_aux+arm_b.n))/(max_xb(i_aux+arm_b.n)- ...
          min_xb(i_aux+arm_b.n)) ;
    end

    % Implement the MLP neural networks
    y_mlp_a(:,i) = mlp_a_r2(x_mlp_a(:,i)'); % MLP - arm A
    y_mlp_b(:,i) = mlp_b_r2(x_mlp_b(:,i)'); % MLP - arm B
    Res(:,i) = [d_mlp_a(:,i)-y_mlp_a(:,i); ...
        d_mlp_b(:,i)-y_mlp_b(:,i)]; % residual vector

    for i_aux = 1:arm_a.n,
        if abs(dqa_est(i_aux,i))< fdi.threshold_lck
            dqa_rbf_aux(i_aux,i) = 1;
        end
    end
    for i_aux = 1:arm_b.n,
        if abs(dqb_est(i_aux,i))< fdi.threshold_lck
            dqb_rbf_aux(i_aux,i) = 1;
        end
    end
    x_rbf(:,i)=[Res(:,i);dqa_rbf_aux(:,i);dqb_rbf_aux(:,i)];

    % Implement the RBF neural network
    y_rbf(:,i) = rbf_r5(x_rbf(:,i)'); % RBFN
```

```
        % Fault Criteria
        if ( (i>3) & (fdi.enable==1) )
            for i_aux=1:2*(arm_a.n+arm_b.n)
                if y_rbf(i_aux,i)==max(y_rbf(:,i)) & ...
                    y_rbf(i_aux,i-1)==max(y_rbf(:,i-1)) & ...
                    y_rbf(i_aux,i-2)==max(y_rbf(:,i-2)) & ...
                    y_rbf(i_aux,i-3)==max(y_rbf(:,i-3)),
                    if ( i_aux < (arm_a.n+arm_b.n+1) )
                      Fault_fsw_fdi(i_aux,i) = 1;
                    else
                      Fault_lck_fdi(i_aux-arm_a.n-arm_b.n,i) = 1;
                    end
                end
            end
        end
    end
    ...
```

File: uarm_reconfig.m

```
    % Free-swinging joint fault
    if (sum(Fault_fsw_fdi(:,i))>0)
        for i_aux=1:(arm_a.n+arm_b.n)
            if (Fault_fsw_fdi(i_aux,i) == 1)
                if i_aux<(arm_a.n+1)
                    fdi.reconfiguration_joint_a = i_aux;
                else
                    fdi.reconfiguration_joint_b = i_aux-arm_a.n;
                end
            end
        end
        fdi.faultisolated = 1;
        fdi.reset_squeeze_cont = 1;

        cont.fault = 'FSwg';
    ...

    end

    % Locked joint fault
    if (sum(Fault_lck_fdi(:,i))>0)
        for i_aux=1:(arm_a.n+arm_b.n)
            if (Fault_lck_fdi(i_aux,i) == 1)
                if i_aux<(arm_a.n+1)
                    fdi.reconfiguration_joint_a = i_aux;
                else
                    fdi.reconfiguration_joint_b = i_aux-arm_a.n;
                end
            end
        end
```

```
        fdi.faultisolated = 1;
        %fdi.reset_squeeze_cont = 1;

        cont.fault = 'FLck';

    ...

    end

    ...
```

8.6.2 FDI Results

To implement the fault tolerant framework, two MLPs (one for each robot) are utilized, each one with 12 inputs representing the three joint positions, three joint velocities, three joint torques, and force vector at the end-effector of arm i; 37 neurons in the hidden layer; and 3 outputs representing the estimation of the three joint velocities. The MLPs were trained with 3,250 patterns obtained in 50 trajectories with random initial and final positions of the fault-free cooperative system. The RBFN has 12 inputs and 13 outputs (six FSJFs, six LJFs, and normal operation) and it was trained with 2506 patterns obtained in 240 trajectories of the cooperative system with FSJFs and LJFs at different joints and 20 without fault. The parameters of the FDI system are $d = 4$ samples, $\gamma_{p1} = \gamma_{p2} = 0.05$, $\gamma_{v1} = \gamma_{v2} = 1.5$, and $\delta_i = 4 \times 10^{-3}$.

The FDI system was tested with 360 trajectories of the cooperative system with FSJFs and LJFs at different joints and 15 without faults. The desired trajectories are divided in three sets: the first one considers a load with mass equal to 0.45 kg; the second set considers the mass of load equal to 0.025 kg and different trajectories of set 1; finally, the third set considers the same trajectories of set 2, with mass equal to set 1. The results are summarized in Table 8.1.

Both strategies discussed in Sect. 8.3, namely, reconfiguration of the system starting with zero velocities and with the current velocities, were tested. In the latter case, the new desired trajectory is a third order polynomial with initial velocities of the load equal to the current ones. In the real system, reconfiguration starting with the current velocities should not be applied in cases where the velocities of the load are high, or the resulting desired trajectory may require joint positions outside of their physical limits.

Figures 8.3, 8.4 and 8.5 show the results of the real system when an FSJF at joint 1 of arm 1 is artificially introduced at $t = 1$ s. When the fault is isolated, at $t = 3.8$ s, the desired trajectories and control laws are reconfigured. One can observe that the time necessary to detect the fault in this trajectory (2.8 s) is higher than the mean time to detection (MTD) for the test sets presented in Table 8.1. In

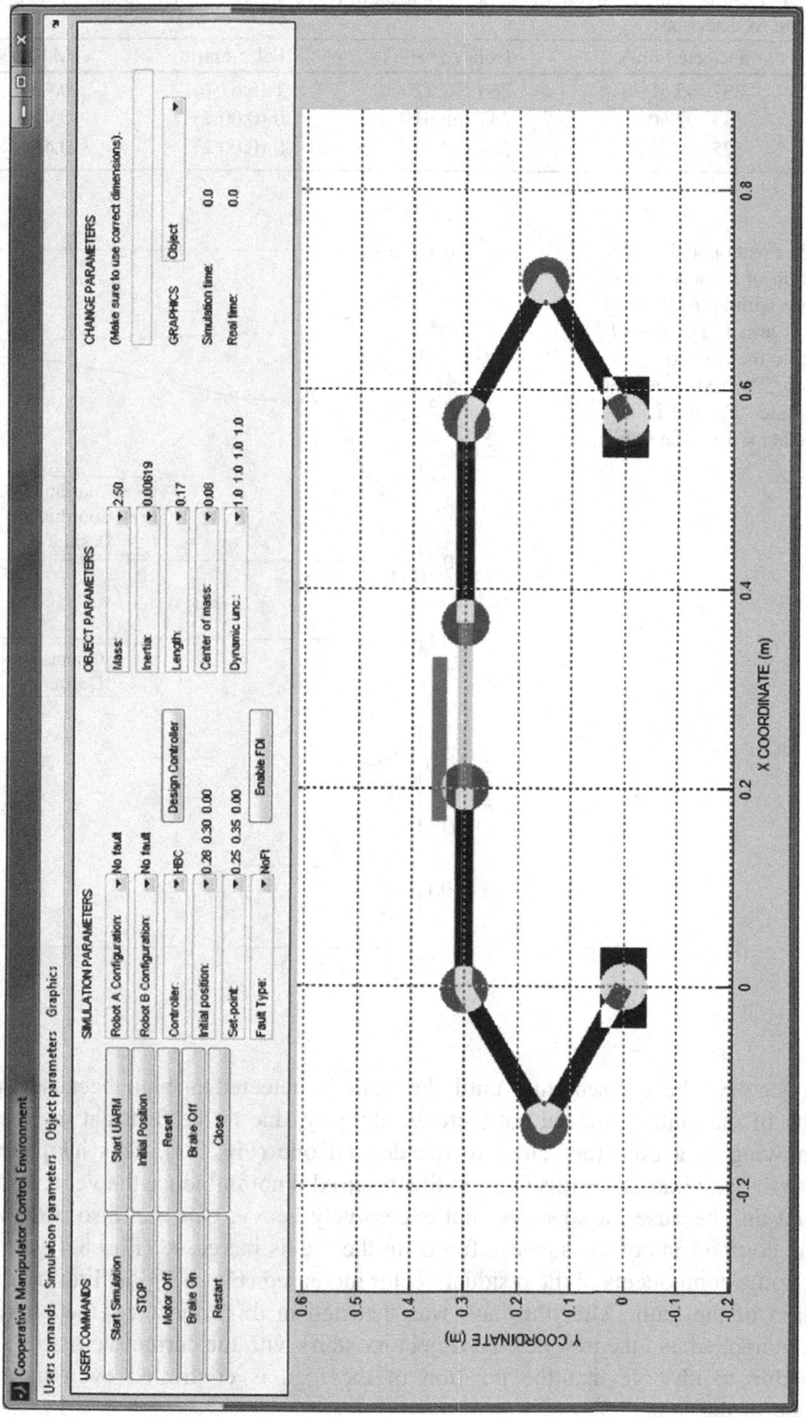

Fig. 8.2 Cooperative Manipulator Control Environment

Table 8.1 Results of the FDI system applied to a team of two cooperative manipulators (MTD = mean time to detection)

Set	Detected faults	Isolated faults	False alarms	MTD (s)
1	337 (93.61%)	260 (72.22%)	1 (6.67%)	0.469
2	333 (92.50%)	247 (68.61%)	0 (0.00%)	0.419
3	325 (90.28%)	268 (74.44%)	0 (0.00%)	0.458

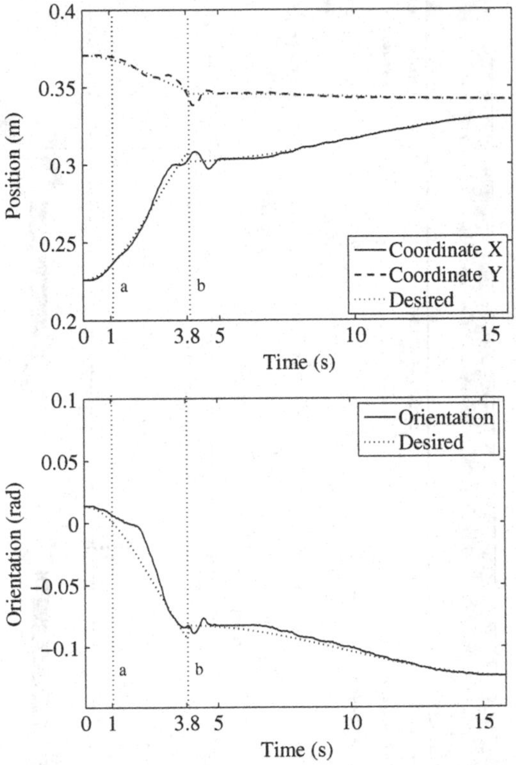

Fig. 8.3 Position and orientation of the load in a trajectory with an FSJF in joint 1 of arm 1. The *dotted lines* show the desired trajectory. The FSJF starts at $t = 1$ s (line "a") and is detected at $t = 3.8$ s (line "b")

this trajectory, the elapsed time until the fault is detected is high because the velocity of the faulty joint did not increase abruptly, due to the fact that the load kept moving in a trajectory close to the desired one (Fig. 8.3). This happened because the gravitational terms (zero in this setup) did not influence the velocity of the joint and because the load was not excessively heavy. One can also observe that the components of the squeeze forces in the y axis increased (Fig. 8.4). As a result, some components of the residual vector increased (Fig. 8.5), facilitating the detection of the fault. After the fault was detected in this trajectory, the brakes were not applied and the new desired trajectory starts with the current velocities. It is possible to observe that the position of the load is controlled even in the presence of the fault.

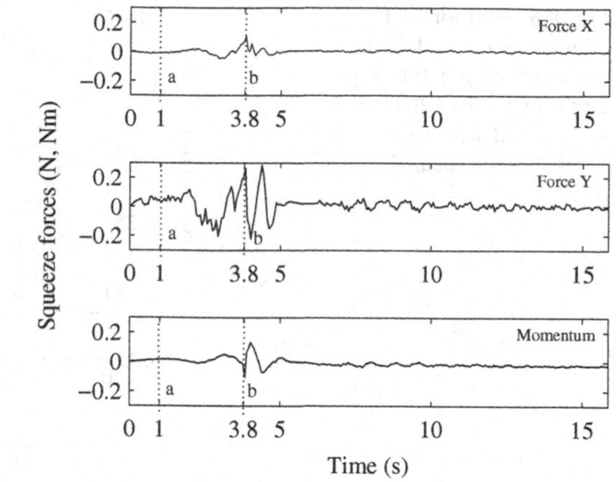

Fig. 8.4 Squeeze forces in the trajectory in Fig. 8.3. The torque component of the squeeze forces is not controlled after the control reconfiguration

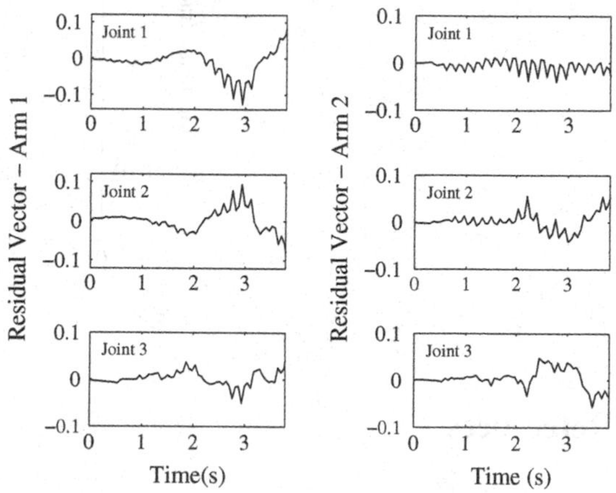

Fig. 8.5 Components of the residual vector in the trajectory in Fig. 8.3. The components are shown from $t = 0$ s to $t = 3.8$ s, when the fault is detected

Figure 8.6 shows the trajectories of the cooperative manipulator system where a JPF at joint 2 of arm 1 is artificially introduced at $t = 1.0$ s and isolated at $t = 1.05$ s. After the fault is detected, the controller ignores the corresponding joint position measurement and utilizes the estimate produced by (8.6).

The performance of the fault detection and isolation system can be improved with additional post-fault tests. When a fault is detected, the brakes can be activated and tests can be performed to verify if the fault was correctly isolated. For example, if a locked joint fault is declared, the controller can try to move this joint in order to confirm the fault isolation. Similar tests can be made to confirm the isolation of other faults. This strategy can also be used to isolate multiple simultaneous faults, which can be detected but cannot be correctly isolated by the fault detection and isolation system presented here. When multiple faults are considered, additional tests to isolate all faults must be made in all joints after the detection of a fault.

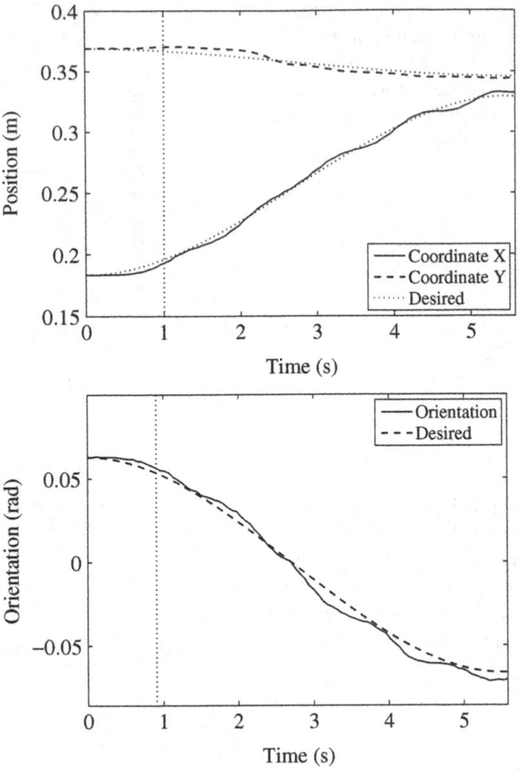

Fig. 8.6 Position and orientation of the load in a trajectory with a JPF in joint 2 of arm 1. The JPF started at $t = 1$ s and was detected at $t = 1.05$ s (*vertical dotted line*)

References

1. Bonitz R, Hsia T (1996) Internal force-based impedance control for cooperating manipulators. IEEE Trans Robot Autom 12(1):78–89
2. Carignan CR, Akin DL (1988) Cooperative control of two arms in the transport of an inertial load in zero gravity. IEEE Trans Robot Autom 4(4):414–419
3. Dhillon BS, Fashandi ARM (1997) Robotic systems probabilistic analysis. Microeletron Reliab 37(2):211–224
4. English JD, Maciejewski AA (1998) Fault tolerance for kinematically redundant manipulators: anticipating free-swinging joint failures. IEEE Trans Robot Autom 14(4): 566–575
5. Goel M, Maciejewski AA, Balakrishnan V (2004) Analyzing unidentified locked-joint failures in kinematically redundant manipulators. J Robot Systs 22(1):15–29
6. Hassan M, Notash L (2004) Analysis of active joint failure in parallel robot manipulators. J Mech Design 126(6):959–968
7. Hassan M, Notash L (2005) Design modification of parallel manipulators for optimum fault tolerance to joint jam. Mech Mach Theory 40(5):559–577
8. Jatta F, Legnani G, Visioli A, Ziliani G (2006) On the use of velocity feedback in hybrid force/velocity control of industrial manipulators. Control Eng Pract 14(9):1045–1055
9. Liu YH, Xu Y, Bergerman M (1999) Cooperation control of multiple manipulators with passive joints. IEEE Transa Robot Autom 15(2):258–267

10. Luh JYS, Zheng YF (1987) Constrained relations between two coordinated industrial robots for robot motion control. Int J Robot Res 6(3):60–70
11. McIntyre ML, Dixon WE, Dawson DM, Walker ID (2005) Fault identification for robot manipulators. IEEE Trans Robot Autom 21(5):1028–1034
12. Miyagi PE, Riascos LAM (2006) Modeling and analysis of fault-tolerant systems for machining operations based on Petri nets. Control Eng Pract 14(4):397–408
13. Monteverde V, Tosunoglu S (1999) Development and application of a fault tolerance measure for serial and parallel robotic structures. Int J Modeling Simulat 19(1):45–51
14. Schneider H, Frank PM (1996) Observer-based supervision and fault-detection in robots using nonlinear and fuzzy logic residual evaluation. IEEE Trans Control Syst Technol 4(3):274–282
15. Terra MH, Tinós R (2001) Fault detection and isolation in robotic manipulators via neural networks—a comparison among three architectures for residual analysis. J Robot Syst 18(7):357–374
16. Tinós R, Terra MH, Bergerman M (2001) Fault detection and isolation in cooperative manipulators via artificial neural networks. In: Proceedings of IEEE Conference on Control Applications, Mexico City, Mexico, pp 988–1003
17. Tinós R, Terra MH, Ishihara JY (2006) Motion and force control of cooperative robotic manipulators with passive joints. IEEE Trans Control Syst Technol 14(4):725–734
18. Vemuri AT, Polycarpou MM (2004) A methodology for fault diagnosis in robotic systems using neural networks. Robotica 22(4):419–438
19. Visinsky ML, Cavallaro JR, Walker ID (1994) Robotic fault detection and fault tolerance: a survey. Reliab Eng Syst Safe 46:139–158
20. Visinsky ML, Cavallaro JR, Walker ID (1995) A dynamic fault tolerance framework for remote robots. IEEE Trans Robot Autom 11(4):477–490
21. Vukobratovic M, Tuneski A (1998) Mathematical model of multiple manipulators: cooperative compliant manipulation on dynamical environments. Mech Mach Theory 33(8):1211–1239
22. Wen T, Kreutz-Delgado K (1992) Motion and force control for multiple robotic manipulators. Automatica 28(4):729–743

Chapter 9
Robust Control of Cooperative Manipulators

9.1 Introduction

In this chapter we deal with the problem of robust position control for cooperative manipulators rigidly connected to an undeformable load. Design paradigms to solve force/position control problems have been established in the literature to improve the performance of the cooperation. In [8] a control strategy for cooperative manipulators was proposed based on the independence of the position and force controls. The applied force between the manipulator end-effectors and the object is decomposed into motion force and squeeze force, which must be controlled. In [7], the hybrid position/force controller proposed in [8] is extended to underactuated cooperative manipulators.

Cooperative manipulators, like any electromechanical system, are subject to parametric uncertainties and external disturbances. In [4], a semi-decentralized adaptive fuzzy controller with $\mathcal{H}_\infty$-performance is developed for fully-actuated cooperative manipulators. In that work, the dynamic model is derived using the order reduction procedure proposed in [5] for constraint manipulators; only simulation results are presented to validate the controller's performance.

In this chapter, two nonlinear $\mathcal{H}_\infty$ control techniques based on centralized control strategies are evaluated for underactuated cooperative manipulators: $\mathcal{H}_\infty$ control for linear parameter varying (LPV) systems [9] and $\mathcal{H}_\infty$ control based on game theory [2]. These controllers are applied considering the control strategy proposed in [8], where the squeeze force control is designed independently of the position control. In these cases, the $\mathcal{H}_\infty$ performance index considers only the position control problem.

The adaptive fuzzy-based controller for fully-actuated cooperative manipulators proposed in [4] includes the position and squeeze force errors in the $\mathcal{H}_\infty$ performance index. Following a similar approach, in this chapter we present a neural network-based $\mathcal{H}_\infty$ control for fully-actuated and underactuated manipulators [6]. We assume that a nominal dynamic model is available for the neural network to approximate the model. As in [4], the $\mathcal{H}_\infty$ performance index includes the position

A. A. G. Siqueira et al., *Robust Control of Robots*,
DOI: 10.1007/978-0-85729-898-0_9, © Springer-Verlag London Limited 2011

and squeeze force errors, which guarantees an overall disturbance rejection. Furthermore, for underactuated manipulators, a practical solution is presented for the squeeze force control. In this case, only some components of the squeeze force can be controlled and constraints are imposed on the components that are not controlled. The approach guarantees asymptotic convergence of the motion tracking errors in spite of parametric uncertainties and external disturbances. Experimental results obtained with two cooperative UARM systems illustrate their efficiency.

This chapter is organized as follows: Sects. 9.2 and 9.3 present the dynamic equations for fully-actuated and underactuated cooperative manipulators, respectively. Section 9.4 presents the squeeze force control problem. Section 9.5 presents the necessary dynamic model formulation to implement $\mathcal{H}_\infty$ controllers via quasi-LPV representation and via game theory. Section 9.6 develops the neural network-based adaptive $\mathcal{H}_\infty$ control approach for cooperative manipulators. Section 9.7 concludes the chapter with design procedures, experimental results, and comparative studies.

9.2 Fully-Actuated Cooperative Manipulators

Consider a system composed of m fully-actuated cooperative manipulators, each one with n degrees of freedom. Let $q_i \in \Re^n$ be the vector of generalized coordinates of manipulator i and $x_o \in \Re^n$ the position and orientation of the load, which is rigidly connected to the end-effectors of the individual manipulators. As discussed in Chap. 7, this configuration generates a geometric constraint of the form $x_o = \phi_i(q_i)$, for $i = 1, 2, \ldots, m$. The corresponding velocity constraints are given by Eq. 7.2, which can be written as:

$$\dot{q}_i = -J_i^{-1}(q_i)J_{o_i}(x_o)\dot{x}_o,$$

for $i = 1, 2, \ldots, m$, where $J_i(q_i)$ is the Jacobian matrix from joint velocities to end-effector velocities of arm i and $J_{o_i}(x_o)$ is the Jacobian matrix from load velocities to end-effector velocities of arm i. We assume that $J_i^{-1}(q_i)$ is well-defined. The kinematic constraints can be expressed by:

$$\dot{\theta} = \begin{bmatrix} I_n \\ -J^{-1}(q)J_o(x_o) \end{bmatrix} \dot{x}_o \equiv B(x_o)\dot{x}_o, \tag{9.1}$$

where $\theta = [x_o^T q_1^T \cdots q_m^T]^T$, $q = [q_1^T \cdots q_m^T]^T$, $J_o(x_o) = [J_{o_1}^T(x_o) \cdots J_{o_m}^T(x_o)]^T$, and

$$J(q) = \begin{bmatrix} J_1(q_1) & \cdots & 0 \\ \vdots & \ddots & \vdots \\ 0 & \cdots & J_m(q_m) \end{bmatrix}.$$

The dynamic equation of the load is given by:

$$M_o(x_o)\ddot{x}_o + C_o(x_o, \dot{x}_o)\dot{x}_o + g_o(x_o) = J_o^T(x_o)h, \tag{9.2}$$

where $M_o(x_o)$ is the inertia matrix, $C_o(x_o, \dot{x}_o)$ is the Coriolis and centripetal matrix, $g_o(x_o)$ is the gravitational torque vector, and $h = [h_i^T \cdots h_m^T]^T$, with $h_i \in \Re^n$, is the vector of forces applied by manipulator i on the load.

The dynamic equation of manipulator i is given by:

$$M_i(q_i)\ddot{q}_i + C_i(q_i, \dot{q}_i)\dot{q}_i + g_i(q_i) = \tau_i + J_i^T(q_i)h_i, \tag{9.3}$$

where $M_i(q_i)$ is the inertia matrix, $C_i(q_i, \dot{q}_i)$ is the Coriolis and centripetal matrix, $g_i(q_i)$ is the gravitational torque vector, and τ_i is the applied torque vector. The dynamic equation of the overall cooperative system can be written as:

$$M(\theta)\ddot{\theta} + C(\theta, \dot{\theta})\dot{\theta} + g(\theta) = \begin{bmatrix} 0 \\ \tau \end{bmatrix} + \begin{bmatrix} J_o^T(x_o) \\ J^T(q) \end{bmatrix} h, \tag{9.4}$$

where $g(\theta) = [g_o(x_o)^T g_1(q_1)^T \cdots g_m^T(q_m)]^T$, and $\tau = [\tau_1^T \cdots \tau_m^T]^T$,

$$M(\theta) = \begin{bmatrix} M_o(x_o) & 0 & \cdots & 0 \\ 0 & M_1(q_1) & \cdots & 0 \\ \vdots & \vdots & \ddots & \vdots \\ 0 & 0 & \cdots & M_m(q_m) \end{bmatrix}, \quad \text{and}$$

$$C(\theta, \dot{\theta}) = \begin{bmatrix} C_o(x_o, \dot{x}_o) & 0 & \cdots & 0 \\ 0 & C_1(q_1, \dot{q}_1) & \cdots & 0 \\ \vdots & \vdots & \ddots & \vdots \\ 0 & 0 & \cdots & C_m(q_m, \dot{q}_m) \end{bmatrix}.$$

The projection of the applied force h on the frame fixed on the center of mass of the load ($h_o = J_{oq}^T h$) can be orthogonally decomposed as:

$$h_o = h_{os} + h_{om}, \tag{9.5}$$

where $h_{os} \in X_s$ is the vector of squeeze forces and $h_{om} \in X_m$ is the vector of motion forces (see Chap. 7 for details). Considering this decomposition of forces, the dynamic equation of the cooperative manipulator (9.4) can be represented as:

$$M(\theta)\ddot{\theta} + C(\theta, \dot{\theta})\dot{\theta} + g(\theta) = \tau_v + \overline{W}^T(\theta)h_{os}, \tag{9.6}$$

where τ_v is an auxiliary control input given by

$$\tau_v = \begin{bmatrix} W^T h_{om} \\ \tau + J^T(q)J_{oq}^{-T}(x_o)h_{om} \end{bmatrix},$$

$\overline{W}(\theta) = [W\, J_{oq}^{-1}(x_o)\, J(q)]$ is a Jacobian matrix, and $W^T = [I_n\, I_n \cdots I_n]$. If the auxiliary control input is partitioned in two vectors, $\tau_{v1} = W^T h_{om}$ and $\tau_{v2} = \tau + J^T(q)\, J_{oq}^{-T}(x_o) h_{om}$, the applied torque vector can be computed by:

$$\tau = \tau_{v2} - J^T(q) J_{oq}^{-T}(x_o)(W^T)^{\#}\tau_{v1}, \tag{9.7}$$

where $(W^T)^{\#} = W(W^T W)^{-1}$ is the pseudo-inverse of W^T. The motion force is given by $h_{om} = (W^T)^{\#}\tau_{v1}$. Hence, the control problem is to find an auxiliary control τ_v that guarantees stability and robustness against disturbances.

Considering the kinematic constraints (9.1) and multiplying the dynamic equation of the cooperative manipulator (9.6) by $B^T(x_o)$ (to eliminate the squeeze force term, since $B^T(x_o)\overline{W}^T(x_o) = 0$) we obtain:

$$\overline{M}(x_o)\ddot{x}_o + \overline{C}(x_o, \dot{x}_o)\dot{x}_o + \overline{g}(x_o) = \overline{\tau}_v, \tag{9.8}$$

where $\overline{M}(x_o) = B^T(x_o)M(x_o)B(x_o)$, $\overline{g}(x_o) = B^T(x_o)g(x_o)$, $\overline{\tau}_v = B^T(x_o)\tau_v$, and $\overline{C}(x_o, \dot{x}_o) = B^T(x_o)M(x_o)\dot{B}(x_o) + B^T(x_o)C(x_o, \dot{x}_o)B(x_o)$.

9.3 Underactuated Cooperative Manipulators

We assume now that the joints of the cooperative manipulator include n_a active joints (with actuators) and n_p passive joins (without actuators or whose actuators failed). The kinematic constraints (9.1) can be rewritten as:

$$\dot{\widetilde{\theta}} = \begin{bmatrix} I_n \\ -J_{AP}^{-1}(q)J_o(x_o) \end{bmatrix} \dot{x}_o \equiv \widetilde{B}(x_o)\dot{x}_o, \tag{9.9}$$

where $\widetilde{\theta} = [x_o^T\, q_a^T\, q_p^T]^T$, $q_a \in \Re^{n_a}$ is vector of active joint positions, $q_p \in \Re^{n_p}$ is the vector of passive joint positions and $J_{AP}(q)$ is a Jacobian matrix generated from the orthogonal permutation matrix P_{AP} [7]. Therefore, if

$$\widetilde{q} = [q_a^T\, q_p^T]^T = P_{AP}[q_1^T\, q_2^T \cdots q_m^T]^T,$$

then

$$J_{AP}(q) = [J_a(q_a)\, J_p(q_p)] = J(q)P_{AP}.$$

The dynamic equation of the underactuated cooperative manipulator system is given by:

$$\widetilde{M}(\widetilde{\theta})\ddot{\widetilde{\theta}} + \widetilde{C}(\widetilde{\theta}, \dot{\widetilde{\theta}})\dot{\widetilde{\theta}} + \widetilde{g}(\widetilde{\theta}) = \begin{bmatrix} 0 \\ \tau_a \\ 0 \end{bmatrix} + \begin{bmatrix} J_o^T(x_o) \\ J_a^T(q_a) \\ J_p^T(q_p) \end{bmatrix} h, \tag{9.10}$$

where $\widetilde{g}(\widetilde{\theta}) = [g_o(x_o)^T g_{AP}(\widetilde{q})^T]^T$, $g_{AP} = P_{AP}[g_1^T(q_1) \cdots g_2^T(q_2)]^T$,

$$\widetilde{M}(\widetilde{\theta}) = \begin{bmatrix} M_o(x_o) & 0 \\ 0 & M_{AP}(\widetilde{q}) \end{bmatrix}, \quad \widetilde{C}(\widetilde{\theta}, \dot{\widetilde{\theta}}) = \begin{bmatrix} C_o(x_o, \dot{x}_o) & 0 \\ 0 & C_{AP}(\widetilde{q}, \dot{\widetilde{q}}) \end{bmatrix},$$

$$M_{AP}(\widetilde{q}) = P_{AP} \begin{bmatrix} M_1(q_1) & \cdots & 0 \\ \vdots & \ddots & \vdots \\ 0 & \cdots & M_m(q_m) \end{bmatrix} P_{AP}^T, \text{ and}$$

$$C_{AP}(\widetilde{q}, \dot{\widetilde{q}}) = P_{AP} \begin{bmatrix} C_1(q_1, \dot{q}_1) & \cdots & 0 \\ \vdots & \ddots & \vdots \\ 0 & \cdots & C_m(q_m, \dot{q}_m) \end{bmatrix} P_{AP}^T.$$

Taking into account the orthogonal decomposition of the applied forces' projection, we can rewrite Eq. 9.10 as:

$$\widetilde{M}(\widetilde{\theta})\ddot{\widetilde{\theta}} + \widetilde{C}(\widetilde{\theta}, \dot{\widetilde{\theta}})\dot{\widetilde{\theta}} + \widetilde{g}(\widetilde{\theta}) = \tau_v + \widetilde{W}^T(\widetilde{\theta})h_{os}, \tag{9.11}$$

where τ_v is the auxiliary control input given by

$$\tau_v = \begin{bmatrix} W^T h_{om} \\ \tau_a + J_a^T(q_a)J_{oq}^{-T}(x_o)h_{om} \\ J_p^T(q_p)J_{oq}^{-T}(x_o)h_{om} \end{bmatrix},$$

and $\widetilde{W}(\widetilde{\theta}) = [W \;\; J_{oq}^{-1}(x_o)J_a(q_a) \;\; J_{oq}^{-1}(x_o)J_p(q_p)]$ is a Jacobian matrix. If the auxiliary control input is partitioned in three vectors, $\tau_{v1} = W^T(x_o)h_{om}$, $\tau_{v2} = \tau_a + J_a^T(q_a)J_{oq}^{-T}(x_o)h_{om}$, and $\tau_{v3} = J_p^T(q_p)J_{oq}^{-T}(x_o)h_{om}$, the applied torque in the active joints can be computed as:

$$\tau_a = \tau_{v2} - J_a^T(q_a)J_{oq}^{-T}(x_o)\begin{bmatrix} W^T \\ J_p^T(q_p)J_{oq}^{-T}(x_o) \end{bmatrix}^{\#}\begin{bmatrix} \tau_{v1} \\ \tau_{v3} \end{bmatrix}. \tag{9.12}$$

Considering the kinematic constraints (9.9) and multiplying Eq. 9.11 by $\widetilde{B}^T(x_o)$, the dynamic equation of the underactuated cooperative manipulator is given by:

$$\widetilde{M}(x_o)\ddot{x}_o + \widetilde{C}(x_o, \dot{x}_o)\dot{x}_o + \widetilde{g}(x_o) = \widetilde{\tau}_v, \tag{9.13}$$

where

$$\begin{aligned} \widetilde{M}(x_o) &= \widetilde{B}^T(x_o)\widetilde{M}(\widetilde{\theta})\widetilde{B}(x_o), \\ \widetilde{C}(x_o, \dot{x}_o) &= \widetilde{B}^T(x_o)\Big(\widetilde{M}(\widetilde{\theta})\dot{\widetilde{B}}(x_o) + \widetilde{C}(\widetilde{\theta}, \dot{\widetilde{\theta}})\widetilde{B}(x_o)\Big), \\ \widetilde{g}(x_o) &= \widetilde{B}^T(x_o)\widetilde{g}(\widetilde{\theta}), \\ \widetilde{\tau}_v &= \widetilde{B}^T(x_o)\tau_v. \end{aligned}$$

9.4 Motion and Squeeze Force Control

From the control paradigm for cooperative manipulators introduced in [8], the position and squeeze force control problems can be decomposed and solved independently. In this case, the applied torque can be computed by:

$$\tau = \tau_{mg} + \tau_s,$$

where τ_{mg} are torques generated by the position controller and τ_s are torques generated by the squeeze force controller. In this chapter, τ_{mg} is given by (9.7) if all joints of every manipulator are actuated. When at least one joint in the entire cooperative system is underactuated, we use $\tau_{mg} = P_{AP}^{-1}[\tau_a^T\ 0]^T$, with τ_a given by (9.12). In Sect. 9.5, the dynamic equations (9.8) and (9.13) are used to design robust controllers for position control of cooperative manipulators, taking into account parametric uncertainties and external disturbances in the manipulator and in the object.

In the case of fully-actuated cooperative manipulators, Wen and Kreutz–Delgado [8] propose an integral squeeze force controller with torque given by:

$$\tau_s = D^T(\theta)\left[h_{sc}^d + K_i \int_{t_0}^{t} (h_{sc}^d - h_{sc})dt\right], \tag{9.14}$$

where h_{sc} is the vector of squeeze forces unaffected by the motion, h_{sc}^d is the vector of desired squeeze forces, K_i is a positive definite matrix, and

$$D(\theta) = \begin{bmatrix} J_{o_1}^{-1}(x_o)J_1(q_1) & \cdots & 0 \\ \vdots & \ddots & \vdots \\ 0 & \cdots & J_{o_m}^{-1}(x_o)J_m(q_m) \end{bmatrix}.$$

In the case of underactuated cooperative manipulators, Eq. 9.14 can be partitioned as:

$$\begin{bmatrix} \tau_{sa} \\ 0 \end{bmatrix} = \begin{bmatrix} D_{sa}^T(\widetilde{\theta}) \\ D_{sp}^T(\widetilde{\theta}) \end{bmatrix} W_c^T \gamma_s, \tag{9.15}$$

where $[D_{sa}(\widetilde{\theta})\ D_{sp}(\widetilde{\theta})] = D(\theta)P_{AP}$, W_c^T is the full rank matrix that projects the null space of W^T, i.e., $Im(W_c^T) = X_S$, and γ_s is the squeeze force control variable. Note that n_p constraints are imposed in the components of γ_s since it is not possible to apply torque on the passive joints ($\tau_{sp} = 0$). As the manipulators considered here are nonredundant, not all components of γ_s can be independently controlled.

The vector γ_s is partitioned in two parts: the independently-controlled components $\gamma_{sc} \in \Re^{n_e}$, where $n_e = n(m-1) - n_p$ if $n_a > n$ and $n_e = 0$ if $n_a < n$; and the uncontrolled components $\gamma_{sn} \in \Re^{n_p}$. Note that if $n_a < n$, none of the components of γ_s can be controlled. The squeeze force controller is given by:

$$\gamma_{sc} = \Gamma_{sc}^d + K_{is} \int_{t_0}^{t} (\Gamma_{sc}^d - \Gamma_{sc}) dt, \tag{9.16}$$

where Γ_{sc}^d is the desired value for γ_{sc}, Γ_{sc} is the vector of measured squeeze forces, and K_{is} is a positive definite matrix. γ_{sn} is computed from the constraints as a function of γ_{sc}. The torque applied in the active joints is related to the squeeze force control (Eq. 9.15) as:

$$\tau_{sa} = D_{sa}^T(\widetilde{\theta}) W_c^T \gamma_{sc}. \tag{9.17}$$

9.5 Nonlinear $\mathcal{H}_\infty$ Control

In this section the nonlinear $\mathcal{H}_\infty$ control solutions presented in Chaps. 3 and 5 for fully-actuated and single underactuated manipulators, respectively, are applied to cooperative manipulators. The main idea is to consider the reduced-order models (9.8) and (9.13), which represent the dynamic model of cooperative manipulators in a similar way as it is commonly used in the representation of single manipulators.

9.5.1 Control Design via Quasi-LPV Representation

We develop in this section the quasi-LPV representations of fully actuated and underactuated cooperative manipulators based on the following dynamic equation:

$$\widehat{M}_0(x_o)\ddot{x}_o + \widehat{C}_0(x_o, \dot{x}_o)\dot{x}_o + \widehat{g}_0(x_o) + \widehat{\tau}_d = \widehat{\tau}_v, \tag{9.18}$$

where $\widehat{M}_0(x_o) = \overline{M}_0(x_o)$, $\widehat{C}_0(x_o, \dot{x}_o) = \overline{C}_0(x_o, \dot{x}_o)$, $\widehat{g}_0(x_o) = \overline{g}_0(x_o)$, and $\widehat{\tau}_v = \overline{\tau}_v$ if the manipulators are fully-actuated (Eq. 9.8); or $\widehat{M}_0(x_o) = \widetilde{M}_0(x_o)$, $\widehat{C}_0(x_o, \dot{x}_o) = \widetilde{C}_0(x_o, \dot{x}_o)$, $\widehat{g}_0(x_o) = \widetilde{g}_0(x_o)$, and $\widehat{\tau}_v = \widetilde{\tau}_v$ if any of the manipulators is underactuated (Eq. 9.13). The index 0 indicates nominal values for the matrices and vectors. $\widehat{\tau}_d$ is the vector of parametric uncertainties and external disturbances in the manipulators and load. The state tracking error is defined as:

$$\widetilde{x} = \begin{bmatrix} x_o - x_o^d \\ \dot{x}_o - \dot{x}_o^d \end{bmatrix} = \begin{bmatrix} \dot{\widetilde{x}}_o \\ \widetilde{x}_o \end{bmatrix}, \tag{9.19}$$

where x_o^d and $\dot{x}_o^d \in \Re^n$ are the reference trajectory and desired velocity of the load, respectively. The quasi-LPV representations of cooperative manipulators are found using Eqs. 9.18 and 9.19:

$$\dot{\widetilde{x}} = A(x_o, \dot{x}_o)\widetilde{x} + Bu + Bw, \tag{9.20}$$

with $w = \widehat{M}_0^{-1}(x_o)\widetilde{\tau}_d$, $B = [0 \;\; I_n]^T$, and

$$A(x_o, \dot{x}_o) = \begin{bmatrix} 0 & I_n \\ 0 & -\widehat{M}_0^{-1}(x_o)\widehat{C}_0(x_o, \dot{x}_o) \end{bmatrix}.$$

From this equation, the variable $\widehat{\tau}_v$ can be represented as:

$$\widehat{\tau}_v = \widehat{M}_0(x_o)(\ddot{x}_o^d + u) + \widehat{C}_0(x_o, \dot{x}_o)\dot{x}_o^d + \widehat{g}_0(x_o).$$

Although the matrices $\widehat{M}_0(x_o)$ and $\widehat{C}_0(x_o, \dot{x}_o)$ explicitly depend on the load position, x_o, we can consider it as function of the position error, $\widetilde{x}_o$. Hence, Eq. 9.20 is a quasi-LPV representation of fully-actuated and underactuated cooperative manipulators.

9.5.2 Control Design via Game Theory

In this section, game theory is used to solve the $\mathcal{H}_\infty$ control problem of cooperative manipulators. The solution is based on the results presented in Chap. 3. From (9.19), after the state transformation given by:

$$\widetilde{z} = \begin{bmatrix} \widetilde{z}_1 \\ \widetilde{z}_2 \end{bmatrix} = T_0\widetilde{x} = \begin{bmatrix} T_1 \\ T_2 \end{bmatrix}\widetilde{x} = \begin{bmatrix} I & 0 \\ T_{11} & T_{12} \end{bmatrix}\begin{bmatrix} \widetilde{x}_o \\ \dot{\widetilde{x}}_o \end{bmatrix}, \tag{9.21}$$

where T_{11}, $T_{12} \in \Re^{n\times n}$ are constant matrices to be determined, the dynamic equation of the state tracking error becomes

$$\dot{\widetilde{x}} = A_T(\widetilde{x}, t)\widetilde{x} + B_T(\widetilde{x}, t)u + B_T(\widetilde{x}, t)w, \tag{9.22}$$

with $w = \widehat{M}_0(x_o)T_{12}\widehat{M}_0^{-1}(x_o)\widetilde{\tau}_d$,

$$A_T(\widetilde{x}, t) = T_0^{-1}\begin{bmatrix} -T_{12}^{-1}T_{11} & T_{12}^{-1} \\ 0 & -\widehat{M}_0^{-1}(x_o)\widehat{C}_0(x_o, \dot{x}_o) \end{bmatrix}T_0,$$

$$B_T(\widetilde{x}, t) = T_0^{-1}\begin{bmatrix} 0 \\ \widehat{M}_0^{-1}(x_o) \end{bmatrix}.$$

The relationship between the auxiliary control input, $\widehat{\tau}_v$, and the control input, u, is given by:

$$\widehat{\tau}_v = \widehat{M}_0(x_o)\ddot{x}_o^c + \widehat{C}_0(x_o, \dot{x}_o)\dot{x}_o + \widehat{g}_0(x_o), \tag{9.23}$$

with $\ddot{x}_o^c = \ddot{x}_o^d - T_{12}^{-1}T_{11}\dot{\tilde{x}}_o - T_{12}^{-1}\widehat{M}_0^{-1}(x_o)(\widehat{C}_0(x_o,\dot{x}_o)B^T T_0\tilde{x} - u)$. The procedure for finding the solution for the control input u that ensures $\mathcal{H}_\infty$ performance follows the guidelines presented in Chap. 3 for single manipulators.

9.6 Neural Network-Based Adaptive Nonlinear $\mathcal{H}_\infty$ Control

In the previous section we assumed that the dynamic model of the cooperative manipulator system is well-known. In this section we develop another control strategy based on the estimation of the uncertain part of the dynamic model through a neural network-based adaptive control law. Consider again the kinematic constraints (9.1) and the dynamic model for fully-actuated cooperative manipulators (9.6):

$$M(x_o)B(x_o)\ddot{x}_o + \big(M(x_o)\dot{B}(x_o) + C(x_o,\dot{x}_o)B(x_o)\big)\dot{x}_o + g(x_o) + \tau_d = \tau_v + \overline{W}^T(x_o)h_{os}. \tag{9.24}$$

Consider also a bounded desired trajectory for the load, $x_o^d \in \Re^n$, and a bounded desired squeeze force, $h_{os}^d \in \Re^{nm}$, and define the following auxiliary variable:

$$q_r = B(x_o)(\Lambda_o e_o + \dot{x}_o^d) - \eta E_2 e_f,$$

where $q_r \in \Re^{n(m+1)}$, $e_o = x_o^d - x_o$, Λ_o is a symmetric positive definite matrix, $\eta > 0$, $E_2 = [0\ I_{nm}]^T \in \Re^{n(m+1)\times nm}$, and $e_f \in \Re^{nm}$ is the output of the following stable filter

$$\dot{e}_f + \eta e_f = -\lambda_f E_2^T \overline{W}^T(x_o)\tilde{h}_{os}, \tag{9.25}$$

with a symmetric positive definite matrix Λ_f and the squeeze force error $\tilde{h}_{os} = h_{os}^d - h_{os}$.

A composite error signal can be defined as:

$$s = q_r - B(x_o)\dot{x}_o = B(x_o)(\Lambda_o e_o + \dot{e}_o) - \eta E_2 e_f,$$

where $s \in \Re^{n(m+1)}$. Another representation of the composite error can be obtained by applying the stable filter (9.25) and defining the error terms:

$$e_1 \equiv [e_o^T\ \ (J^T(x_o)J_{oq}^{-T}(x_o)\tilde{h}_{os})^T]^T,$$

and $e_2 \equiv [\dot{e}_o^T\ \ \dot{e}_f^T]^T$:

$$\begin{aligned} s &= L\Big[E_1(\Lambda_o e_o + \dot{e}_o) + E_2\Big(\lambda_f E_2^T \overline{A}^T(x_o)\tilde{h}_{os} + \dot{e}_f\Big)\Big] \\ &= L(\Lambda e_1 + e_2), \end{aligned} \tag{9.26}$$

where $L = [B(x_o)\ E_2] \in \Re^{n(m+1)\times n(m+1)}$, $E_1 = [I_n\ \ 0]^T \in \Re^{n(m+1)\times n}$ and $\Lambda = diag[\Lambda_o, \Lambda_f]$. From (9.24) and (9.26), the error dynamic model is given by:

$$\begin{aligned} M(x_o)\dot{s} &= M(x_o)\dot{q}_r - M(x_o)\big(B(x_o)\ddot{x}_o + \dot{B}(x_o)\dot{x}_o\big) \\ &= -C(x_o,\dot{x}_o)s + F_0(x_e) + \Delta F(x_e) - \overline{W}^T(x_o)h_{os} + \tau_d - \tau_v, \end{aligned} \tag{9.27}$$

where $x_e = [x_o^T\ \dot{x}_o^T\ q_r^T\ \dot{q}_r^T]^T$, $F_0(x_e) = M_0(x_o)\dot{q}_r + C_0(x_o,\dot{x}_o)q_r + g_0(x_o)$, and $\Delta F(x_e) = \Delta M(x_o)\dot{q}_r + \Delta C(x_o,\dot{x}_o)q_r + \Delta g(x_o)$.

The terms $M_0(x_o)$, $C_0(x_o,\dot{x}_o)$, and $g_0(x_o)$ represent the nominal values of the matrices $M(x_o)$, $C(x_o,\dot{x}_o)$, and $g(x_o)$, respectively. The parametric uncertainties are represented by $\Delta M(x_o)$, $\Delta C(x_o,\dot{x}_o)$, and $\Delta g(x_o)$.

Following the development of the neural network-based adaptive control strategy presented in Chap. 4, a set of k ($k = 1,\ldots,n$) neural networks $\Delta F(x_e,\Theta_k)$ is used to approximate the uncertain term $\Delta F(x_e)$ in (9.27). Each neural network is composed of nonlinear neurons in the hidden layer and linear neurons in the input and output layers, with adjustable parameters Θ_k in the output layers. The single-output neural networks are of the form:

$$\Delta F_k(x_e,\Theta_k) = \sum_{i=1}^{p_k} H\left(\sum_{j=1}^{5n} w_{ij}^k x_{ej} + m_i^k\right)\Theta_{ki} = \xi_k^T\Theta_k, \tag{9.28}$$

with

$$\xi_k = \begin{bmatrix} H\left(\sum_{j=1}^{5n} w_{1j}^k x_{ej} + m_1^k\right) \\ \vdots \\ H\left(\sum_{j=1}^{5n} w_{p_kj}^k x_{ej} + m_{p_k}^k\right) \end{bmatrix}, \quad \Theta_k = \begin{bmatrix} \Theta_{k1} \\ \vdots \\ \Theta_{kp_k} \end{bmatrix},$$

where p_k is the number of neurons in the hidden layer. The weights w_{ij}^k and the biases m_i^k for $1 \le i \le p_k$, $1 \le j \le 5n$ and $1 \le k \le n$ are assumed to be constant and specified by the designer. $H(.)$ is selected to be a hyperbolic tangent function. The complete neural network can be represented as:

$$\begin{aligned} \Delta F(x_e,\Theta) &= \begin{bmatrix} \Delta F_1(x_e,\Theta_1) \\ \vdots \\ \Delta F_n(x_e,\Theta_n) \end{bmatrix} = \begin{bmatrix} \xi_1^T\Theta_1 \\ \vdots \\ \xi_n^T\Theta_n \end{bmatrix} \\ &= \begin{bmatrix} \xi_1^T & 0 & \ldots & 0 \\ 0 & \xi_2^T & \vdots & 0 \\ \vdots & \vdots & \ddots & \vdots \\ 0 & 0 & \ldots & \xi_n^T \end{bmatrix} \begin{bmatrix} \Theta_1 \\ \Theta_2 \\ \vdots \\ \Theta_n \end{bmatrix} = \Xi\Theta. \end{aligned} \tag{9.29}$$

To guarantee the stability of the $\mathcal{H}_\infty$ controller we consider two fundamental assumptions proposed in [1]:

(1) There exists an optimal parameter value $\Theta^{\star} \in \Omega_\Theta$ such that $\Delta F(x_e, \Theta^{\star})$ approximates $\Delta F(x_e)$ as closely as possible, where Ω_Θ is a pre-assigned constraint region.
(2) The approximation error $\delta F(x_e) = \Delta F(x_e) - \Delta F(x_e, \Theta^{\star})$ must be bounded by a state-dependent function; that is, there exists a function $k(x_e) > 0$ such that $|(\delta F(x_e)_i| \leq k(x_e)$, for all $1 \leq i \leq n$.

Based on these assumptions, the error dynamics can be rewritten as:

$$\begin{aligned} M(x_o)\dot{s} = & - C(x_o, \dot{x}_o)s + F_0(x_e) + \Delta F(x_e, \Theta^{\star}) + \delta F(x_e) \\ & - \overline{W}^T(x_o)h_{os} + \tau_d - \tau_v. \end{aligned} \tag{9.30}$$

The nonlinear $\mathcal{H}_\infty$ adaptive neural network control problem for cooperative manipulators can then be formulated as follows: given a level of attenuation γ, find an auxiliary control input τ_v such that the following $\mathcal{H}_\infty$ performance index is achieved:

$$\begin{aligned} \int_0^T s^T \Psi s dt = & \int_0^T e^T Q e dt \\ & \leq \frac{1}{2} s^T(0) M(x_o(0)) s(0) + \frac{1}{2} \rho e_f^T(0) \Lambda_f^{-1} e_f(0) \\ & + \tilde{\Theta}^T(0) Z \tilde{\Theta}(0) + \gamma^2 \int_0^T \tau_d^T \tau_d dt, \end{aligned} \tag{9.31}$$

where $e = \left[e_I^T \, e_2^T\right]^T$, Z is a symmetric positive definite matrix, $\tilde{\Theta} = \Theta^* - \Theta$ denotes the neural network parameter estimation error, τ_d is a square-integrable torque disturbance ($\tau_d \in L_2$), and

$$Q = \begin{bmatrix} \Lambda^T \Psi \Lambda & \Lambda^T \Psi \\ \Psi \Lambda & \Psi \end{bmatrix}, \tag{9.32}$$

with $\Psi = \Psi^T > 0$ and $\Lambda = \text{diag}[\Lambda_o, \Lambda_f]$.

Theorem *Consider a cooperative manipulator described by Eq. 9.4. If the control law is defined as:*

$$\dot{\Theta} = Proj[S^{-T} \Xi^T s], \tag{9.33}$$

$$\tau_v = F_0(x_e) + Ks + \Xi\Theta - \overline{W}^T(x_o)h_{os} + \frac{\rho}{\eta}\overline{W}^T(x_o)\tilde{h}_{os} + \tau_s, \tag{9.34}$$

with $\tau_s = k(x_e)sgn(s)$ *and*

$$Proj[S^{-T}\Xi^T s] = \begin{cases} S^{-T}\Xi^T s, & \text{if } \Theta^T\Theta \leq M \text{ or } (\Theta^T\Theta > M \\ & \text{and } \Theta^T S^{-T}\Xi^T s \leq 0), \\ S^{-T}\Xi^T s - \frac{(\Theta^T\Theta - M)\Theta^T S^{-T}\Xi^T s}{\delta\Theta^T\Theta}\Theta, & \text{otherwise} \end{cases}$$

and $K = diag[K_o, K_1, \ldots, K_m]$, *with symmetric positive definite matrices* K_o, $K_1, \ldots, K_m$, *and* $Proj[S^{-T}\Xi^T s]$ *is a projection algorithm, then the closed-loop error system satisfies:*

(1) e_o, $\dot{e}_o$, *and* $\ddot{e}_o \in L_\infty$, e_f, $\dot{e}_f$, *and* $\tilde{h}_{os} \in L_\infty$, *and* $\Theta \in \Omega_\Theta$.
(2) *The* $\mathcal{H}_\infty$ *performance index (9.31) is achieved if* K_i *is selected as* $K_i = P_i + (1/4\gamma)I_n$ *with symmetric positive definite matrix* P_i.
(3) *If* $d \in L_2$, *then the motion tracking errors* e_o, $\dot{e}_o$ *converge to zero as* $t \to \infty$.

Proof Consider the Lyapunov function:

$$V = \frac{1}{2}s^T M s + \frac{1}{2}\rho e_f^T \Lambda_f^{-1} e_f + \frac{1}{2}\tilde{\Theta}^T S\tilde{\Theta}.$$

The time derivative of V along the error dynamics (9.27) and control law (9.34) is given by:

$$\begin{aligned} \dot{V} &= s^T M\dot{s} + \frac{1}{2}s^T\dot{M}s + \rho e_f^T\Lambda_f^{-1}\dot{e}_f + \dot{\tilde{\Theta}}^T S\tilde{\Theta} \\ &= s^T\left(-Cs - Ks + \Xi\tilde{\Theta} + \delta F(x_e) - \frac{\rho}{\eta}\overline{W}^T(x_o)\tilde{h}_{os}\right) + \frac{1}{2}s^T\dot{M}s \\ &\quad + \rho e_f^T\Lambda_f^{-1}\left(-\eta e_f - \Lambda_f E_2^T\overline{W}^T(x_o)\tilde{h}_{os}\right) + \dot{\tilde{\Theta}}^T S\tilde{\Theta} - s^T\tau_s + s^T\tau_d. \end{aligned}$$

Since $-(\rho/\eta)s^T\overline{W}^T(x_o)\tilde{h}_{os} - \rho e_f^T E_2^T\overline{W}^T(x_o)\tilde{h}_{os} = 0$ and $(1/2\dot{M} - C)$ is a skew-symmetric matrix, $\dot{V}$ becomes:

$$\dot{V} = -s^T Ks + s^T\Xi\tilde{\Theta} + s^T\delta F(x_e) - \eta\rho e_f^T\Lambda_f^{-1}e_f + \dot{\tilde{\Theta}}^T S\tilde{\Theta} - s^T\tau_s + s^T\tau_d. \quad (9.35)$$

From the definition of the update law (9.33), where the projection algorithm is used, we can show that:

$$\dot{\tilde{\Theta}}^T S\tilde{\Theta} + s^T\Xi\tilde{\Theta} \leq 0, \quad (9.36)$$

and $\Theta(t) \in \Omega_\Theta$ for all $t \geq 0$ if $\Theta(0) \in \Omega_0$, with $\Omega_0 = \{\Theta : \Theta^T\Theta \leq M\}$. Taking into account the control law τ_s and assumption (2), the following inequality can be guaranteed:

$$s^T(-\tau_s+\delta F(x_e)) = s^T(-k(x_e)sgn(s)-\delta F(x_e)) \\ \leq -k(x_e)\sum_{i=1}^{n(m+1)}|s_i| + \sum_{i=1}^{n(m+1)}|(\delta F(x_e))_i||s_i| \leq 0. \quad (9.37)$$

Substituting Eqs. 9.36 and 9.37 into Eq. 9.35, we obtain:

$$\dot{V} \leq -s^T K s - \eta\rho e_f^T \Lambda_f^{-1} e_f + s^T \tau_d. \quad (9.38)$$

According to the inequality $s^T\tau_d \leq (1/\gamma^2 4)s^T s + \gamma^2 \tau_d^T \tau_d$, Eq. 9.38 leads to:

$$\dot{V} \leq -s^T P s + \gamma^2 \tau_d^T \tau_d, \quad (9.39)$$

where the control gains $K_i = P_i + (1/4\gamma^2)I_n$ have been used, and $P = \text{diag}[P_o, P_1, \ldots, P_m]$. Substituting (9.26) in (9.39), we obtain:

$$\dot{V} \leq -e^T \begin{bmatrix} \Lambda^T L^T P L \Lambda & \Lambda^T L^T P L \\ L^T P L \Lambda & L^T P L \end{bmatrix} e + \gamma^2 \tau_d^T \tau_d \\ = -e^T Q e + \gamma^2 \tau_d^T \tau_d. \quad (9.40)$$

Considering $\Psi = L^T P L$, matrix Q satisfies Eq. 9.32 and is symmetric positive semi-definite. By integrating the above inequality, the closed-loop system satisfies the following $\mathcal{H}_\infty$ performance index:

$$\int_0^T e^T Q e dt \leq V(0) - V(t) + \gamma^2 \int_0^T \tau_d^T \tau_d dt \\ \leq \frac{1}{2} s^T(0) M(0) s(0) + \frac{1}{2}\rho e_f^T(0)\Lambda_f^{-1} e_f(0) \\ + \tilde{\Theta}^T(0) Z \tilde{\Theta}(0) + \gamma^2 \int_0^T \tau_d^T \tau_d dt. \quad (9.41)$$

From Eq. 9.39, $\dot{V}$ is convergent, which implies $s \in L_\infty$, and, therefore, that e_o, $\dot{e}_o$, and $e_f \in L_\infty$. The squeeze force error can be expressed as a function of the neural network approximation parameters ($\Delta\hat{M}$, $\Delta\hat{C}$, and $\Delta\hat{g}$) as:

$$\eta \overline{W}(x_o) M^{-1}(x_o)\Big((M_0(x_o) - \Delta\hat{M})E_2\Lambda_f E_2^T + \rho/(\eta^2) I_{n(m+1)}\Big)\overline{W}^T(x_o)\tilde{h}_{os} \\ = z - \eta^2 \overline{W}(x_o) M^{-1}(x_o)(M_0(x_o) - \Delta\hat{M})E_2 e_f + \overline{W}(x_o)\dot{B}(x_o)\dot{x}_o, \quad (9.42)$$

with

$$z = \overline{W}(x_o)M^{-1}(x_o)\Big(-C(x_o,\dot{x}_o)s - Ks - \Delta\hat{C}q_r - \Delta\hat{g} + \tau_d - \tau_s \\ -(M_0(x_o) - \Delta\hat{M})\big(\dot{B}(x_o)(\Lambda_o e_o + \dot{x}_o^d) + B(x_o)(\Lambda_o \dot{e}_o + \ddot{x}_o^d)\big)\Big).$$

Since all terms on the right-hand side of Eq. 9.42 are bounded, a proper value of ρ assures $\tilde{h}_{os}$ is bounded and, therefore, $\dot{e}_f$, $\dot{s}$, and $\ddot{e}_o \in L_\infty$.

If a square-integrable torque disturbance is assumed, i.e., $\tau_d \in L_2$, then $s \in L_2$ by integrating Eq. 9.39. By Barbalat's lemma, $\lim_{t\to\infty} s(t) = 0$, since $s \in L_2$ and $s, \dot{s} \in L_\infty$. Hence, $\lim_{t\to\infty} \dot{e}_o, e_o, e_f = 0$. □

The algorithm $Proj[S^{-T}\Xi^T s]$ was originally defined in [3]. It guarantees that $\Theta(t) \in \Omega_\Theta$ for all t, where $\Omega_\Theta = \{\Theta : \Theta^T\Theta \leq M + \delta\}$, for some $M > 0$ and $\delta > 0$ a pre-assigned constraint region for Θ.

We now consider the underactuated case, where the cooperative manipulator is characterized by n_a active joints and n_p passive joints. Considering the kinematic constraints (9.9), the dynamic equation of the underactuated cooperative manipulator is given by:

$$\begin{aligned}&\widetilde{M}(x_o)\widetilde{B}(x_o)\ddot{x}_o + \left(\widetilde{M}(x_o)\dot{\widetilde{B}}(x_o) + \widetilde{C}(x_o, \dot{x}_o)\widetilde{B}(x_o)\right)\dot{x}_o + \widetilde{g}(x_o) + \tau_d \\ &= \widetilde{\tau}_v + \widetilde{W}^T(x_o)h_{os}.\end{aligned} \tag{9.43}$$

The update and control laws for underactuated cooperative manipulators can now be defined as:

$$\dot{\widetilde{\Theta}} = Proj[\widetilde{S}^{-T}\widetilde{\Xi}^T s],$$

$$\widetilde{\tau}_v = \widetilde{F}_0(x_e) + Ks + \widetilde{\Xi}\widetilde{\Theta} - \widetilde{W}^T(x_o)h_{os} + \frac{\rho}{\eta}\widetilde{W}^T(x_o)\tilde{h}_{os} + \tau_s,$$

where $\widetilde{F}_0(x_e) = \widetilde{M}_0(x_o)\dot{q}_r + \widetilde{C}_0(x_o, \dot{x}_o)q_r + \widetilde{g}_0(x_o)$and $\Delta\widetilde{F}(x_e, \widetilde{\Theta}) = \widetilde{\Xi}\widetilde{\Theta}$ is the neural network used to approximate $\Delta\widetilde{F}(x_e) = \Delta\widetilde{M}(x_o)\dot{q}_r + \Delta\widetilde{C}(x_o, \dot{x}_o)q_r + \Delta\widetilde{g}(x_o)$.

In the fully-actuated cooperative system, n actuated joints are needed to control the components of the motion of the load and $n(m-1)$ actuated joints are utilized to control the squeeze forces. However, when underactuated cooperative manipulators are considered, not all squeeze force components can be controlled since some degrees of actuation have been lost.

Consider the squeeze force error defined in (9.25). The dimension of $\tilde{h}_{os}$ is nm, and since the dimension of X_s is $n(m-1)$, it is possible to write $\tilde{h}_{os} = W_c^T \gamma_s$, where $\gamma_s \in \Re^{n(m-1)}$ and $W_c^T \in \Re^{nm\times n(m-1)}$ is the full-rank matrix that projects the null space of W^T, that is, $\mathrm{Im}(W_c^T) = X_s$. Hence, the $n(m-1)$-dimensional vector γ_s is a variable to be controlled.

The squeeze force error term, $(\rho/\eta)\widetilde{W}^T(x_o)\widetilde{h}_{os}$, can be described as:

$$\begin{bmatrix} \tau_{os} \\ \tau_{as} \\ \tau_{ps} \end{bmatrix} = (\rho/\eta) \begin{bmatrix} W^T \\ J_a^T(x_o)J_{oq}^{-T}(x_o) \\ J_p^T(x_o)J_{oq}^{-T}(x_o) \end{bmatrix} W_c^T \gamma_s,$$

where τ_{os}, τ_{as}, τ_{ps} are the contributions of the squeeze force error in the components of the auxiliary control input $\widetilde{\tau}_v$. By imposing that $\tau_{ps} = 0$, since it is assumed that no actuation occurs at passive joints, n_p constraints are created in the components of γ_s:

$$J_p^T(x_o)J_{oq}^{-T}(x_o)W_c^T \gamma_s = 0. \tag{9.44}$$

Thus, if the manipulators are not kinematically redundant, only n_e components of γ_s can be independently controlled, where

$$n_e = \begin{cases} n(m-1) - n_p, & \text{if } n_a > n, \\ n_e = 0, & \text{otherwise.} \end{cases}$$

The vector γ_s is now partitioned in the independently controlled components $\gamma_{sc} \in \Re^{n_e}$ and in the uncontrolled components $\gamma_{sn} \in \Re^{n_p}$. Note that if $n_a < n$, none of the components of γ_s can be controlled. In the control law implementation, the components of γ_{sn} are computed from the constraints (9.44) as a function of γ_{sc} (see more details in [7]).

9.7 Examples

In this section, the controllers presented in this chapter are implemented and tested in a cooperative manipulator system. We consider two desired trajectories: a straight line for the quasi-LPV and game theory-based controllers, and an arc of circumference for the neural network-based controller.

9.7.1 Design Procedures

The robust controllers can be designed using the Cooperative Manipulator Control Environment (CMCE). The controller can be selected with the menu *Controller* (see Fig. 9.1).

The controllers' gains are loaded in the control environment by executing the file `uarm_gains.m`. To change the controller behavior, the user may click on the button *Design Controller* and select the appropriate controller and fault configuration. Figure 9.2 shows the control design box for the NLH-quasi-LPV controller. The designer can select the following control parameters:

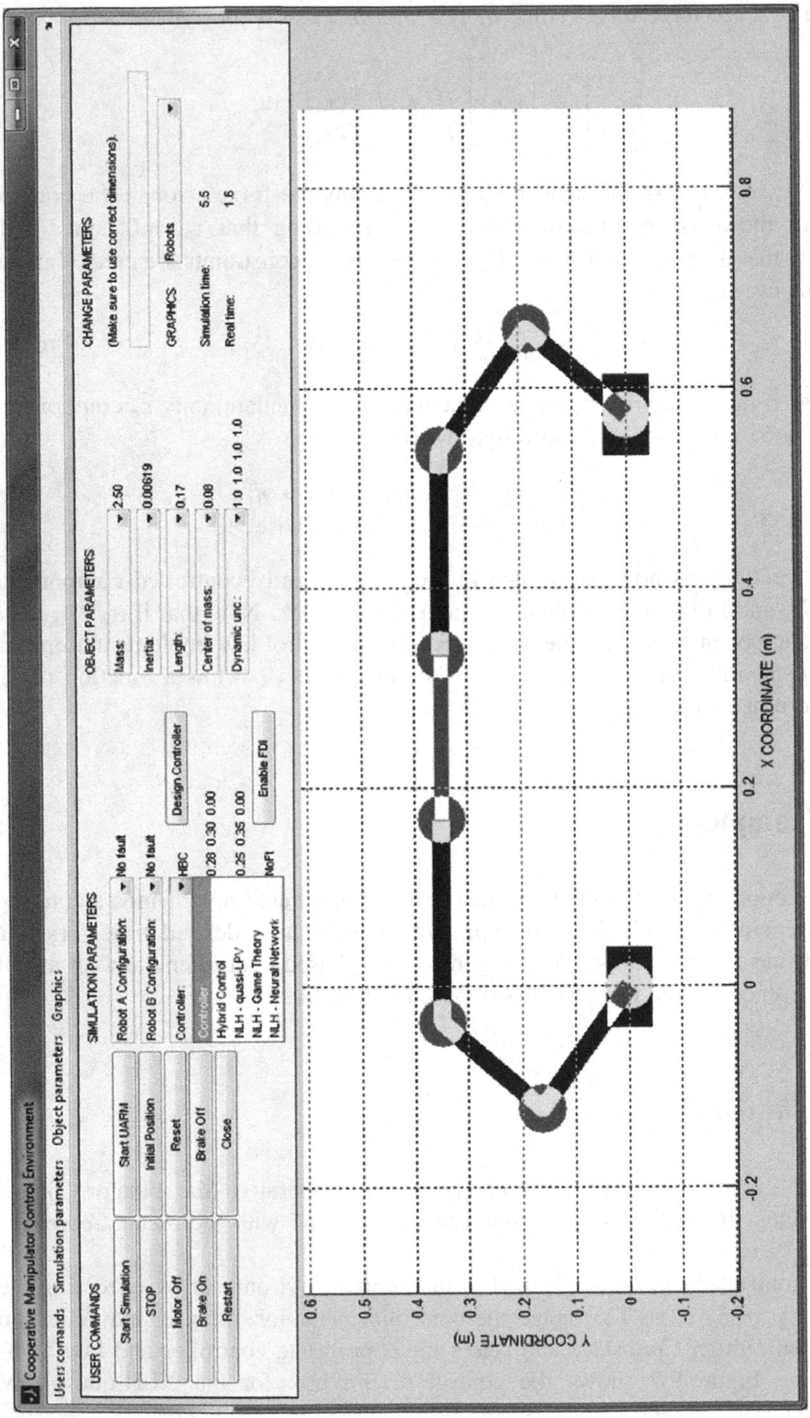

Fig. 9.1 Cooperative Manipulators Control Environment, controller menu

- *L*: Defines the number of points in the grid of the parameter set *P*.
- $X_{\max}$: Defines the maximum absolute value for the first element of the parameter vector ρ, defined as the *X* coordinate of the object. The parameter range is defined as $\rho_1 \in [-X_{\max}, X_{\max}]$.
- $Y_{\max}$: Defines the maximum absolute value for the second element of ρ, the *Y* coordinate of the object. The parameter range is defined as $\rho_2 \in [-Y_{\max}, Y_{\max}]$.
- *phimax*: Defines the maximum value for the third element of ρ, the orientation of the object ϕ. The parameter range is defined as $\rho_3 \in [-\phi_{\max}, \phi_{\max}]$.
- $V_{\max}$: Defines the maximum value for the variation rates of the parameters ρ_i. It is a vector with three entries, one for each parameter.
- *gamma*: Defines the value for the attenuation level γ.
- *Ki*: Defines the squeeze force control parameter.

The following MATLAB® code implements the quasi-LPV and game theory controllers for the free-swinging joint fault configuration.

File: uarm_control.m

```
...

% H-infinity controller - quasi LPV
elseif (all(cont.type == 'LPV'))

    Bap = [eye(n/2); -inv(Jap)*(Jo)];
    dBap = [zeros(n/2);inv(Jap)*dJap*inv(Jap)*(Jo_est) - ...
        inv(J_est)*(dJo_est)];
    AAap = [Jo_est Jap];

    Map = Per*M_est*Per';
    Cap = Per*C_est*Per';

    MBap = Bap'*Map*Bap;
    WBap = inv(MBap);
    CBap = Bap'*Map*dBap + Bap'*Cap*Bap;

    derror = -derror;
    error = -error;
    e(:,i) = error;
    de(:,i)= derror;

    B_2 = [  zeros(3); eye(3) ];
    f1 = 1;
    f2(i) = error(1);
    f3(i) = error(2);
    f4(i) = cos(error(3));
    X_nlh = (f1*X1f + f2(i)*X2f + f3(i)*X3f + f4(i)*X4f);
    u(:,i) = -(B_2'*inv(X_nlh))*[error; derror];

    tauv(:,i) = pinv(Bap')*MBap*ddxo_d(:,i) + ...
       pinv(Bap')*CBap*dxo_d(:,i) + pinv(Bap')*MBap*u(:,i);
```

```
        tau_s = [-Dat_m']*fs_d;

        tau_at = (tauv(4:3+nat,i) - Jat'*inv(Joq_est')* ...
           pinv([A'; Jpa'*inv(Joq_est')])*[tauv(1:3,i); ...
           tauv(3+nat+1:9,i)]) + tau_s;

        tau_pa = zeros(npa,1);

    % H-infinity Controller - Game Theory
    elseif (all(cont.type == 'GTH'))

        Bap = [eye(n/2); -inv(Jap)*(Jo_est)];
        dBap = [zeros(n/2);inv(Jap)*dJap*inv(Jap)*(Jo_est) - ...
           inv(J_est)*(dJo_est)];
        AAap = [Jo_est Jap];

        Map = Per*M_est*Per';
        Cap = Per*C_est*Per';

        MBap = Bap'*Map*Bap;
        WBap = inv(MBap);
        CBap = Bap'*Map*dBap + Bap'*Cap*Bap;
        derror = -derror;
        error = -error;
        e(:,i) = error;
        de(:,i)= derror;

        u(:,i) = -inv(Rc)*Be'*T0*[error;derror];
        ddxo_c(:,i) = ddxo_d(:,i) - inv(T12)*T11*derror - ...
           inv(T12)*inv(MBap)*(CBap*Be'*T0*[error;derror] - ...
           u(:,i));

        tauv(:,i) = pinv(Bap')*MBap*ddxo_c(:,i) + ...
           pinv(Bap')*CBap*dxo(:,i);

        tau_s = [-Dat_m']*fs_d;

        tau_at = (tauv(4:3+nat,i) - Jat'*inv(Joq_est')* ...
           pinv([A'; Jpa'*inv(Joq_est')])*[tauv(1:3,i); ...
           tauv(3+nat+1:9,i)]) + tau_s;
        tau_pa = zeros(npa,1);

        ...
```

9.7.2 Fully-Actuated Configuration

To validate the nonlinear $\mathcal{H}_\infty$ control methods presented in the previous sections we apply them to the underactuated cooperative manipulator shown in Chap. 7, composed of two identical planar underactuated manipulators UARM. The workspace and the coordinate system for the cooperative manipulator are shown in Fig. 9.3; the load parameters are presented in Table 9.1. The kinematic and dynamic parameters of the manipulators can be found in Chap. 1.

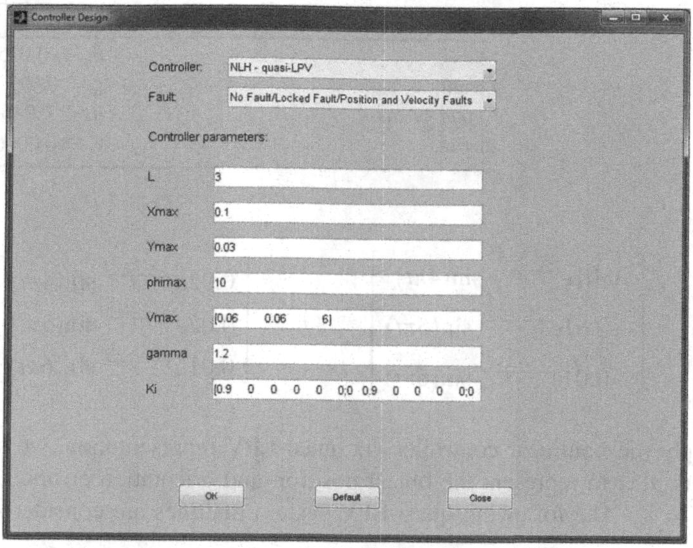

Fig. 9.2 Controller design box, NLH—quasi-LPV control

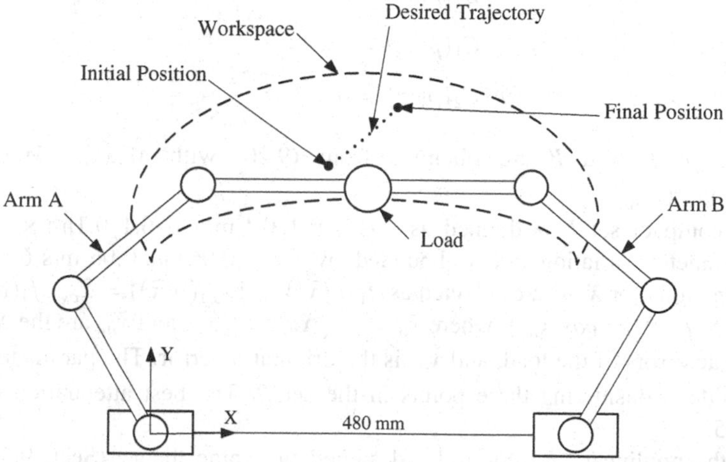

Fig. 9.3 Workspace and coordinate system of the cooperative manipulator UARM and desired trajectory for the model-based controllers

The goal is to move the center of mass of the load along a straight line in the X–Y plane from $x_{\mathrm{o}}(0) = [0.20\mathrm{m}\ 0.35\mathrm{m}\ 0°]^T$ to $x_{\mathrm{o}}^d(T) = [0.25\mathrm{m}\ 0.40\mathrm{m}\ 0°]^T$, where $T = 5.0$ s is the duration of the motion. The reference trajectory $x_{\mathrm{o}}^d(t)$ is generated using a fifth-degree polynomial. The following external disturbances are introduced to verify the robustness of the proposed controllers:

Table 9.1 Load parameters

Parameter	Value
Mass	$m_o = 0.025$ kg
Length	$l_o = 0.092$ m
Center of mass	$a_o = 0.046$ m
Inertia	$I_o = 0.000023\,\text{kg m}^2$

$$\tau_{d_1} = \begin{bmatrix} 0.01e^{-\frac{(t-2.5)^2}{8}} \sin(4\pi t) \\ -0.01e^{-\frac{(t-2.5)^2}{8}} \sin(5\pi t) \\ -0.01e^{-\frac{(t-2.5)^2}{8}} \sin(6\pi t) \end{bmatrix}, \quad \tau_{d_2} = \begin{bmatrix} 0.02e^{-\frac{(t-2.5)^2}{8}} \sin(4\pi t) \\ 0.02e^{-\frac{(t-2.5)^2}{8}} \sin(5\pi t) \\ 0.01e^{-\frac{(t-2.5)^2}{8}} \sin(6\pi t) \end{bmatrix}.$$

To apply the nonlinear controller via quasi-LPV representation, we select the parameters $\rho(\tilde{x})$ to represent the object position and orientation errors, i.e., $k = 3$ and $\rho(\tilde{x}) = \tilde{x}_o$. The following quasi-LPV system matrices are considered:

$$\begin{aligned} A(\rho(x)) &= A(\rho(\tilde{x})), \\ B_1(\rho(x)) &= B, \\ B_2(\rho(x)) &= B, \\ C_1(\rho(x)) &= I_6, \\ C_2(\rho(x)) &= 0, \end{aligned}$$

where $A(\rho(\tilde{x}))$ and B are obtained from (9.20) with $\widehat{M}(x_o) = \overline{M}(x_o)$ and $\widehat{C}(x_o, \dot{x}_o) = \overline{C}(x_o, \dot{x}_o)$.

The compact set P is defined as $\rho \in [-0.1, 0.1]\text{m} \times [-0.1, 0.1]\text{m} \times [-9, 9]^\circ$. The parameter variation rate is bounded by $|\dot{\rho}| \leq [0.06\,\text{m/s}\ 0.06\ \text{m/s}\ 6^\circ/\text{s}]$. The basis functions for $X(\rho)$ are selected as: $f_1(\rho(\tilde{x})) = 1$, $f_2(\rho(\tilde{x})) = \tilde{x}_{oX}$, $f_3(\rho(\tilde{x})) = \tilde{x}_{oY}$ and $f_4(\rho(\tilde{x})) = cos(\tilde{x}_{o_\phi})$, where $\tilde{x}_o = [\tilde{x}_{oX}\ \tilde{x}_{oY}\ \tilde{x}_{o_\phi}]$, $\tilde{x}_{oX}$ and $\tilde{x}_{oY}$ are the X and Y coordinate errors of the load, and $\tilde{x}_{o_\phi}$ is the orientation error. The parameter space was divided considering three points in the set P. The best attenuation level is $\gamma = 1.25$.

For the nonlinear $\mathcal{H}_\infty$ controller designed via game theory (Sect. 9.5.2), the attenuation level is $\gamma = 4.0$. The weighting matrices used are: $Q_1 = 10I_3$, $Q_2 = I_3$, $Q_{12} = 0$, and $R = I_3$. The desired values for the squeeze force are $h_{sc}^d = [0\ \ 0\ \ 0]^T$. The integral gains of the squeeze force controller are $K_i = 0.9I_3$ and $K_{i_s} = 0.9I_3$ for the fully-actuated and underactuated cases, respectively. The resulting Cartesian coordinates and orientation of the object are shown in Figs. 9.4 and 9.5.

Three performance indexes are used to compare the performance of the nonlinear $\mathcal{H}_\infty$ controllers: the $\mathcal{L}_2$ norm of the state vector, $\mathcal{L}_2[\tilde{x}]$, the integral of the

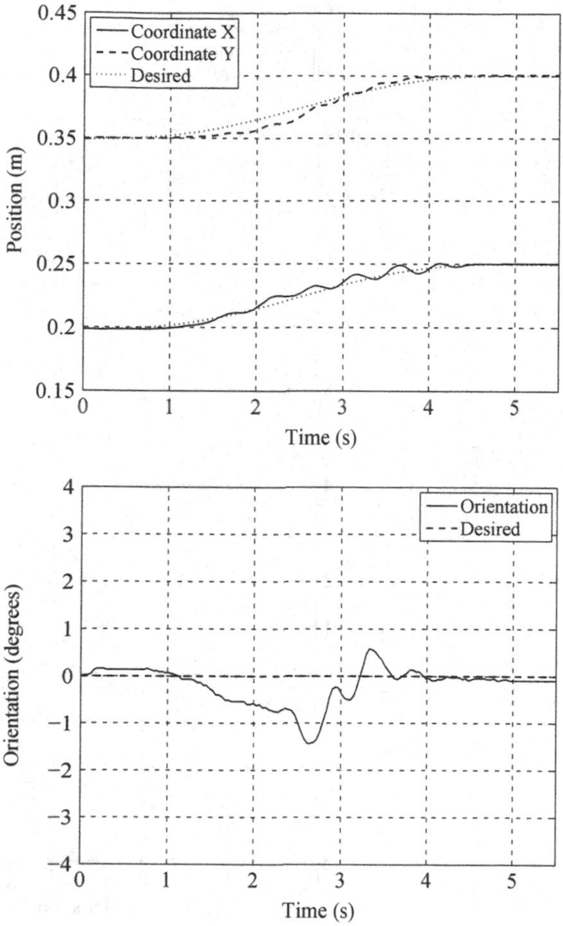

Fig. 9.4 Control of a rigid load by a system of two cooperative fully-actuated manipulators, quasi-LPV formulation

applied torque by the *i-th* joint for both manipulators, $E[\tau]$, and the integral of the squeeze force vector:

$$E[h_{os}] = \sum_{i=1}^{nm} \left(\int_{0}^{t_r} |h_{oS_i}(t)| \mathrm{d}t \right),$$

where t_r is the time it takes for the load to reach the desired position. The results are presented in Tables 9.2 and 9.3 and represent the average of five experiments.

Note that the nonlinear $\mathcal{H}_\infty$ controller designed via game theory presents the lowest trajectory tracking error, $\mathcal{L}_2[\tilde{x}]$, although the energy usage $E[\tau]$ and the squeeze forces $E[h_{os}]$ are higher in comparison to the quasi-LPV formulation.

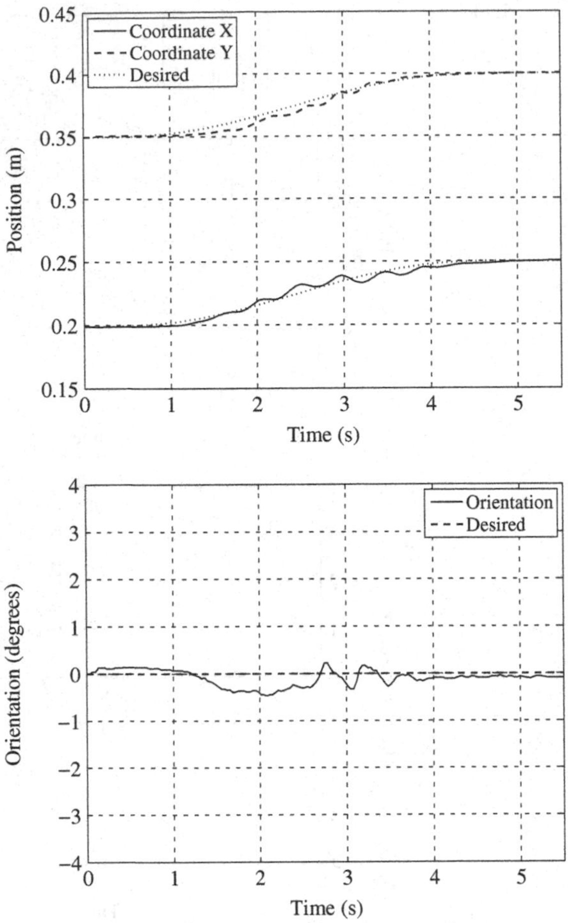

Fig. 9.5 Control of a rigid load by a system of two cooperative fully-actuated manipulators, game theory formulation

Table 9.2 Performance indexes, fully actuated configuration

Control Formulation	$\mathcal{L}_2[\tilde{x}]$	$E[\tau]$ (N m s)	$E[h_{os}]$ (N s)
Quasi-LPV	0.01815	0.8318	0.2193
Game theory	0.01158	1.1200	0.3875

Table 9.3 Performance indexes, underactuated configuration

Nonlinear $\mathcal{H}_\infty$	$\mathcal{L}_2[\tilde{x}]$	$E[\tau]$ (N m s)	$E[h_{os}]$ (N s)
Quasi-LPV	0.0154	0.9976	0.4477
Game theory	0.0103	1.0609	0.3973

9.7.3 Underactuated Configuration

In this section, we assume that joint 1 of manipulator A in Fig. 9.3 is passive. In this case, ($n_e = n(m-1) - n_p = 2$) and therefore only two components of the squeeze force can be controlled independently. We choose to control the X and Y components of the squeeze force but not the component relative to the momentum applied to the load. The desired values for the squeeze force are $\Gamma_{sc}^d = [0 \;\; 0]^T$.

The parameters $\rho(\widetilde{x})$, the variation rate bounds, and the basis functions needed to compute $X(\rho)$ are the same ones used in the fully-actuated case. The quasi-LPV system matrices are also the same except that $\widehat{M}(x_o) = \widetilde{M}(x_o)$ and $\widehat{C}(x_o, \dot{x}_o) = \widetilde{C}(x_o, \dot{x}_o)$. The parameter space was divided considering three points in the set P. The best level of attenuation is $\gamma = 1.25$. The weighting matrices for the nonlinear $\mathcal{H}_\infty$ control via game theory are the same defined for the fully-actuated case. The level of attenuation adopted is $\gamma = 4.0$.

The experimental results are shown in Figs. 9.6 and 9.7, and the performance indexes in Table 9.3. Note that, in this case, the nonlinear $\mathcal{H}_\infty$ controller via game theory presents the lowest trajectory tracking error and squeeze force. The best value for the energy usage is given by the nonlinear $\mathcal{H}_\infty$ controller via quasi-LPV representation.

Figure 9.8 shows, for the quasi-LPV formulation, the squeeze force components when the squeeze force control is applied (continuous line) and when it is not applied (dashed line). It can be observed that only the components of the squeeze force relative to the linear coordinates are controlled and close to the desired values $\Gamma_{sc}^d = 0$. The component of the squeeze force relative to the momentum is not controlled in both cases, as mentioned before.

For the case where the squeeze force is not controlled, the values of $\mathcal{L}_2[\widetilde{x}]$ and $E[\tau]$ are close to the values in Table 9.3. The values of $E[h_{os}]$, however, are on average three times larger than when the squeeze force is controlled.

We ran the same experiment using the hybrid position/force control for underactuated manipulators proposed in [7]. The average performance indexes over five experiments are $\mathcal{L}_2[\widetilde{x}] = 0.0128$, $E[\tau] = 1.7781$, and $E[h_{os}] = 0.5741$. Although in this case $\mathcal{L}_2[\widetilde{x}]$ is lower than that obtained via the quasi-LPV formulation, the values of $E[\tau]$ and $E[h_{os}]$ are approximately 70% and 40% larger, respectively. The conclusion is that, in this case, the robust controllers present practically the same position tracking performance than the hybrid position/force controller, but with less control effort.

9.8 Neural Network-Based Adaptive Controller

We also experimented with controlling the underactuated cooperative manipulator with the nonlinear $\mathcal{H}_\infty$ neural network-based adaptive controller developed in Sect. 9.6. The load parameters, presented in Table 9.4, represent those of a

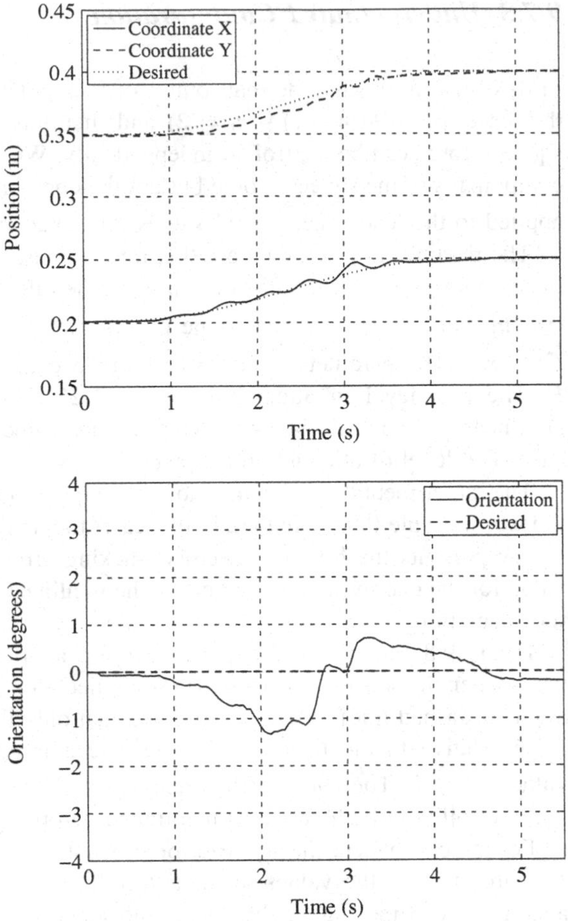

Fig. 9.6 Control of a rigid load by a system of two cooperative underactuated manipulators, quasi-LPV formulation

force-torque sensor attached to the manipulators' end-effectors. Figure 9.9 shows the workspace of the cooperative manipulator and the desired trajectory for the set of experiments presented next.

The desired trajectory is an arc of circle centered at $C = (0.24, 0.08)$ m and with radius $R = 0.26$ m. The arc spans from $x_o(0) = [0.1\,\text{m}\,0.3\,\text{m}\;0°]^T$ to $x_o^d(T) = [0.38\;\;\text{m}\,0.3\,\text{m}\;0°]^T$, where $T = 3.0$ s is the desired duration of the motion. The reference trajectory for the x-axis is a fifth-degree polynomial, and for the y-axis it is defined by the reference arc. The following external disturbances are introduced to verify the robustness of the proposed controllers:

$$\tau_{d_1} = \begin{bmatrix} 0.01e^{-\frac{(t-1.5)^2}{0.5}}\sin(2\pi t) \\ -0.01e^{-\frac{(t-1.5)^2}{0.5}}\sin(2.5\pi t) \\ -0.01e^{-\frac{(t-1.5)^2}{0.5}}\sin(3\pi t) \end{bmatrix}, \quad \tau_{d_2} = \begin{bmatrix} 0.02e^{-\frac{(t-1.5)^2}{0.5}}\sin(2\pi t) \\ 0.02e^{-\frac{(t-1.5)^2}{0.5}}\sin(2.5\pi t) \\ 0.01e^{-\frac{(t-1.5)^2}{0.5}}\sin(3\pi t) \end{bmatrix}.$$

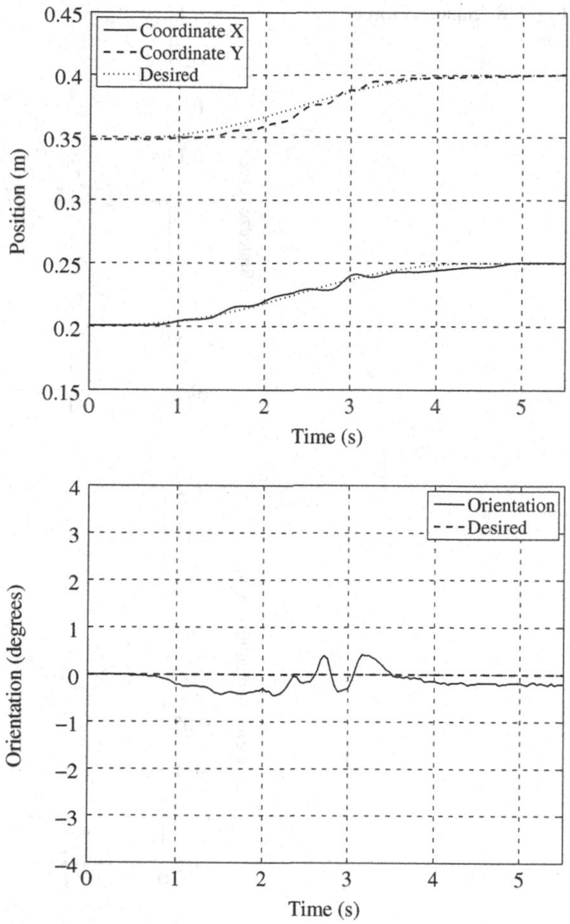

Fig. 9.7 Control of a rigid load by a system of two cooperative underactuated manipulators, game theory formulation

For the fully-actuated configuration, the following control parameters are used: $K = \text{diag}[3.42I_3, 0.38I_6]$, $\Lambda = 0.8I_3$, $\Lambda_f = 0.5I_3$, $\rho = 0.4$, $\eta = 1$, and $S = 50$. The desired values for the squeeze force are $h_{os}^d = [0\ 0\ 0]^T$. To compute the neural network, the following auxiliary variable is defined:

$$\psi = \sum_{i=1}^{n} (x_o)_i + (\dot{x}_o)_i + (q_r)_i + (\dot{q}_r)_i.$$

The matrix Ξ can be computed as:

$$\Xi = diag[\xi_1^T, \xi_2^T, \ldots, \xi_9^T],$$

with $\xi_k = [\xi_{k1}, \ldots, \xi_{k7}]^T$, $\xi_k = (e^{\psi + m_i^k} - e^{-\psi - m_i^k})/(e^{\psi + m_i^k} + e^{-\psi - m_i^k})$, and m_i^k assumes the values $-1.5, -1, -0.5, 0, 0.5, 1, 1.5$, for $i = 1, \ldots, 7$, respectively. Note

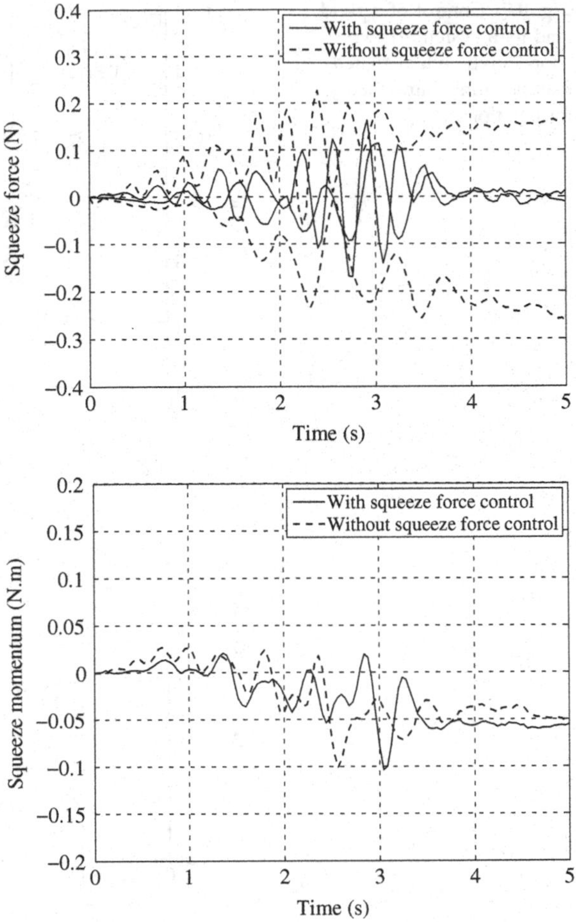

Fig. 9.8 Squeeze force control

Table 9.4 Load parameters

Parameter	Value
Mass	$m_o = 1.45$ kg
Length	$l_o = 0.120$ m
Center of mass	$a_o = 0.060$ m
Inertia	$I_o = 0.0026\ \text{kg m}^2$

that, with these definitions, seven neurons in the hidden layer are selected for the neural networks with the weights w_{ij}^k assuming the value 1. The network parameters Θ are defined as $\Theta = \left[\Theta_1^T, \ldots, \Theta_9^T\right]^T$, with $\Theta_k = \left[\Theta_{k1}\Theta_{k2}\ldots\Theta_{k7}\right]^T$. It is assumed that the approximation error is bounded by the state-dependent function $k(\mathbf{x}_e)$ defined as $k(x_e) = 2\sqrt{\tilde{x}_o^2 + \dot{\tilde{x}}_o^2}$. The resulting Cartesian coordinates and orientation of the load are shown in Fig. 9.10. Table 9.5 compares the values of $\mathcal{L}_2[\tilde{x}]$, $E[\tau]$, and $E[h_{oS}]$ when all three controllers presented in this chapter are applied to the

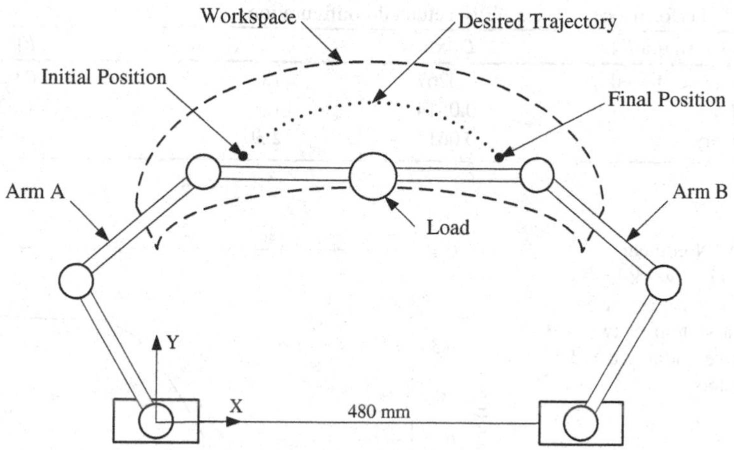

Fig. 9.9 Workspace of the cooperative manipulator system and desired trajectory

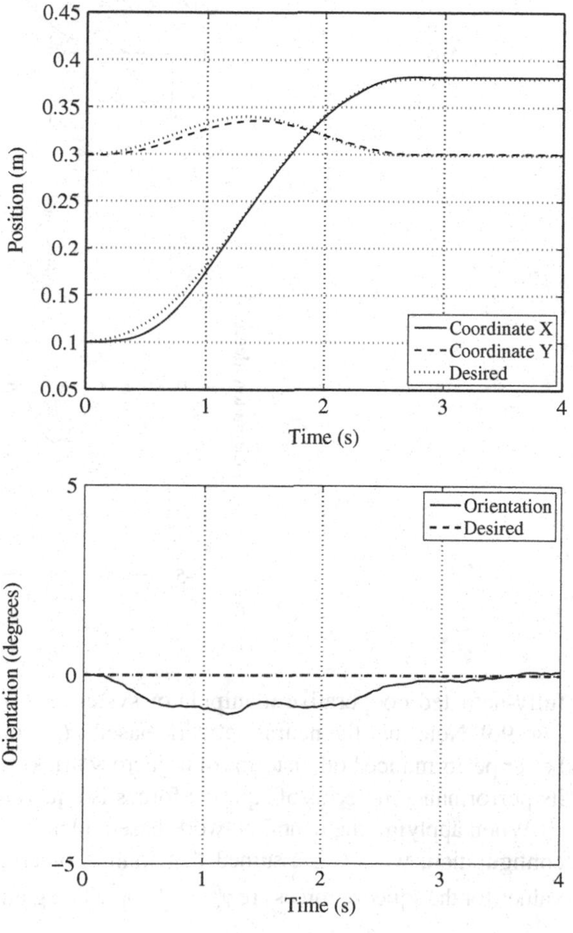

Fig. 9.10 Nonlinear $\mathcal{H}_\infty$ neural network-based adaptive control of a rigid load by a system of two cooperative fully-actuated manipulators

Table 9.5 Performance indexes, fully-actuated configuration

Controller formulation	$\mathcal{L}_2[\tilde{x}]$	$E[\tau]$ (N m s)	$E[h_{os}]$ (N s)
Neural network-based	0.0267	1.09	0.9253
Quasi-LPV	0.0514	1.95	1.02
Game theory	0.0617	2.30	1.11

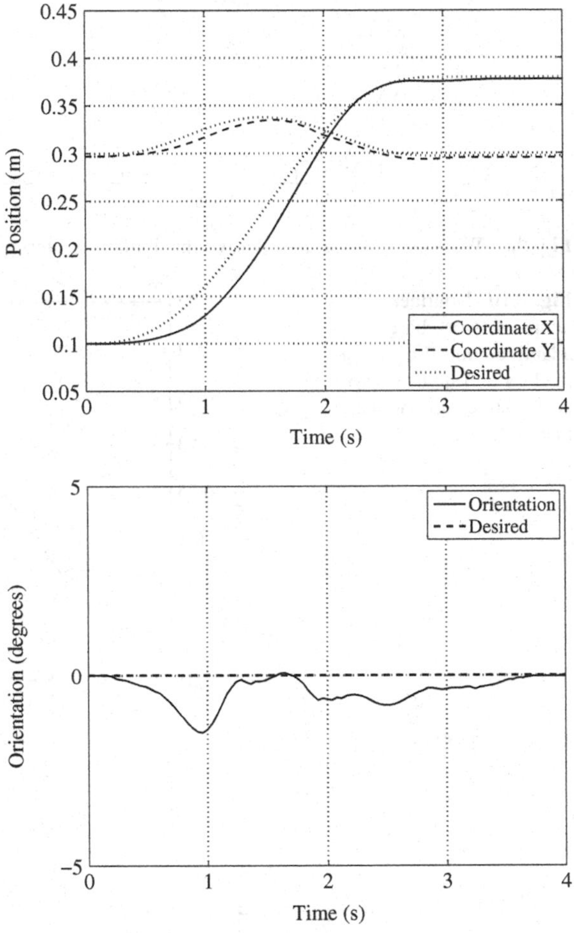

Fig. 9.11 Nonlinear $\mathcal{H}_\infty$ neural network-based adaptive control of a rigid load by a system of two cooperative underactuated manipulators

fully-actuated cooperative manipulator system following the desired trajectory of Fig. 9.9. Note that the neural network-based $\mathcal{H}_\infty$ controller presented significantly better performance both in terms of trajectory tracking error and energy usage, while its performance in terms of squeeze forces is equivalent to the other two.

When applying the neural network-based adaptive controller to the underactuated configuration, we again assumed that joint 1 of arm A is passive and the desired values for the squeeze forces are $\gamma_{sc}^d = [0\ 0]^T$. The gains are $K = \text{diag}[3.42I_3, 0.38I_6]$,

Table 9.6 Performance indexes, underactuated configuration

Controller formulation	$\mathcal{L}_2[\tilde{x}]$	$E[\tau]$ (N m s)	$E[h_{os}]$ (N s)
Neural network-based	0.0462	2.36	1.56
Quasi-LPV	0.0561	2.70	1.61
Game theory	0.0574	2.83	1.65

$\Lambda = 1.5I_3$, $\Lambda_f = 0.4I_3$, $\rho = 0.2$, $\eta = 1$, and $S = 50$. The experimental results are shown in Fig. 9.11. The comparative performance indexes are presented in Table 9.6. Note that, once more, the neural network-based $\mathcal{H}_\infty$ controller presents the best performance in terms of trajectory tracking error and energy usage, while presenting similar performance in terms of squeeze force control.

References

1. Chang YC (2000) Neural network-based $\mathcal{H}_\infty$ tracking control for robotic systems. IEEE Proc Control Theory Appl 147(3):303–311
2. Chen BS, Lee TS, Feng JH (1994) A nonlinear $\mathcal{H}_\infty$ control design in robotic systems under parameter perturbation and external disturbance. Int J Control 59(2):439–461
3. Khalil HK (1996) Adaptive output feedback control of nonlinear systems represented by input-output models. IEEE Trans Autom Control 41(2):177–188
4. Lian KY, Chiu CS, Liu P (2002) Semi-descentralized adaptive fuzzy control for cooperative multirobot systems with $\mathcal{H}_\infty$ motion/internal force tracking performance. IEEE Trans Syst, Man Cybern—Part B: Cybernetics 32(3):269–280
5. McClamroch NH, Wang D (1988) Feedback stabilization and tracking of constrained robots. IEEE Trans Autom Control 33(5):419–426
6. Siqueira AAG, Terra MH (2007) Neural network-based $\mathcal{H}_\infty$ control for fully actuated and underactuated cooperative manipulators. In: Proceedings of the American Control Conference, New York, USA pp 3259–3264
7. Tinós R, Terra MH (2006) Motion and force control of cooperative robotic manipulators with passive joints. IEEE Trans Control Syst Technol 14(4):725–734
8. Wen T, Kreutz-Delgado K (1992) Motion and force control for multiple robotic manipulators. Automatica 28(4):729–743
9. Wu F, Yang XH, Packard A, Becker G (1996) Induced $\mathcal{L}_2$-norm control for LPV systems with bounded parameter variation rates. Int J Robust Nonlinear Control 6(9–10):983–998

Index

A. A. G. Siqueira et al., *Robust Control of Robots*,
DOI: 10.1007/978-0-85729-898-0, © Springer-Verlag London Limited 2011